Contents

Measurement
Comparing Measures 68
Estimating Measures 70
Converting Measures 72
Measurement Calculations 74
12 and 24 Hour Time 76
Time Problems 78
Perimeter and Area 80

Progress Test 3 82

Geometry – Properties of Shapes
2-D Shapes 86
3-D Shapes 88
Lines of Symmetry 90
Angles 92

Geometry – Position and Direction
Translations 94
Coordinates 96
Shapes and Coordinates 98

Statistics
Bar Charts 100
Time Graphs 102
Pictograms 104
Tables 106

Progress Test 4 108

Answers (pull-out) 113

Progress Test Charts (pull-out) 127

ACKNOWLEDGEMENTS

The author and publisher are grateful to the copyright holders for permission to use quoted materials and images.
All illustrations and images are © Shutterstock.com and © HarperCollinsPublishers

Every effort has been made to trace copyright holders and obtain their permission for the use of copyright material. The author and publisher will gladly receive information enabling them to rectify any error or omission in subsequent editions. All facts are correct at time of going to press.

Published by Leckie
An imprint of HarperCollinsPublishers
1 Robroyston Gate, Glasgow, G33 1JN

HarperCollinsPublishers
Macken House, 39/40 Mayor Street Upper, Dublin 1, D01 C9W8, Ireland

© HarperCollinsPublishers Limited 2017

ISBN 978-0-00-866589-0
First published 2017
10 9 8 7 6 5

All rights reserved. No part of this publication may be reproduced, stored in a retrieval system, or transmitted, in any form or by any means, electronic, mechanical, photocopying, recording or otherwise, without the prior permission of Leckie.

Without limiting the exclusive rights of any author, contributor or the publisher of this publication, any unauthorised use of this publication to train generative artificial intelligence (AI) technologies is expressly prohibited. HarperCollins also exercise their rights under Article 4(3) of the Digital Single Market Directive 2019/790 and expressly reserve this publication from the text and data mining exception.

Note for teachers: this book is also available as a downloadable pdf for unlimited school use: ISBN 978-0-00-826387-4

British Library Cataloguing in Publication Data.
A CIP record of this book is available from the British Library.

Series Concept and Development: Michelle I'Anson
Commissioning Editor: Richard Toms
Series Editor: Charlotte Christensen
Author: Tom Hall
Project Manager and Editorial: David Mantovani
Cover Design: Sarah Duxbury
Cover Illustration: Louise Forshaw
Inside Concept Design: Ian Wrigley
Text Design and Layout: Contentra Technologies
Artwork: Collins and Contentra Technologies
Production: Natalia Rebow
Printed in the UK by Ashford Colour Ltd

Starter Test

PS Problem-solving questions

1. Write the next numbers in these sequences.

    a) 16, 20, 24, 28, ___, ___

    b) 350, 400, 450, 500, ___, ___

    2 marks

2. Write seven hundred and sixty in digits. _____

    1 mark

3. Write < or > in the circles to make each number sentence correct.

    a) 658 ◯ 685   b) 804 ◯ 809

    2 marks

4. Circle the number that has a digit with the value of seven tens.

    697   742   871   37   707

    1 mark

5. Write the numbers the arrows point to.

    a) _____   b) _____

    2 marks

6. Calculate:

    a) 5 2 7
       +2 9 4
       _____

    b) 8 0 3
       -2 2 7
       _____

    c) 8 3 8
       -4 8 0
       _____

    3 marks

PS 7. Nia has saved 389 photos on her phone. She deletes 137.

   How many photos does she have left? _____

   1 mark

4

About this book

This workbook contains practice to support your learning in P4/P5 maths.

- Questions split into three levels of increasing difficulty – Challenge 1, Challenge 2 and Challenge 3 – to aid progress.

- Symbol to highlight questions that test problem-solving skills.

- Total marks boxes for each challenge and topic.

  'How am I doing?' checks for self-evaluation.

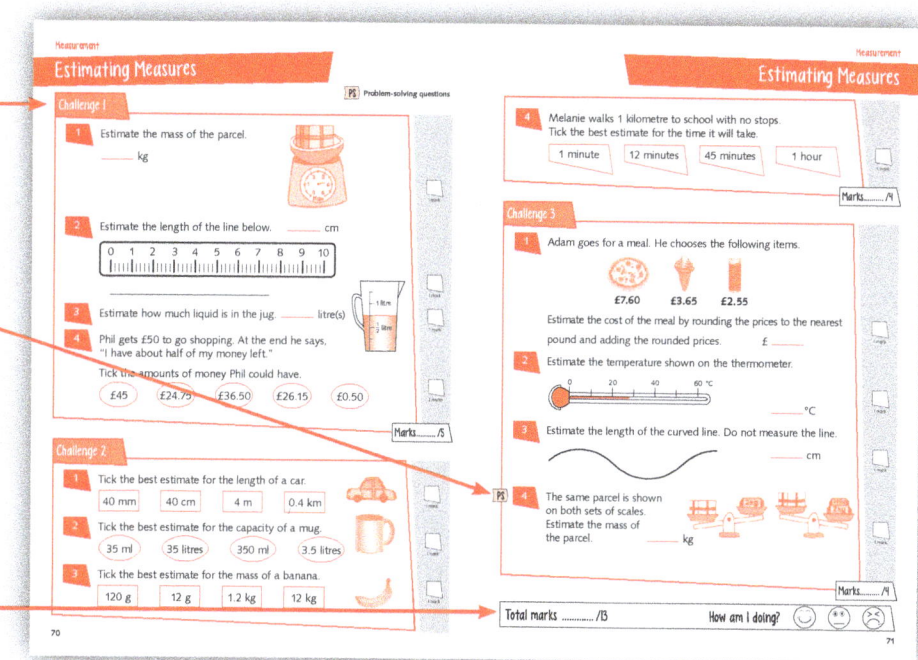

- Starter test recaps skills covered in P4/P5.

- Four progress tests throughout the book, allowing children to revisit the topics and test how well they have remembered the information.

- Progress charts to record results and identify which areas need further revision and practice.

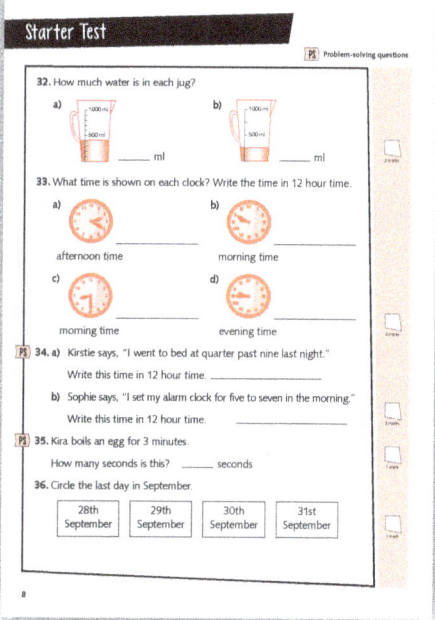
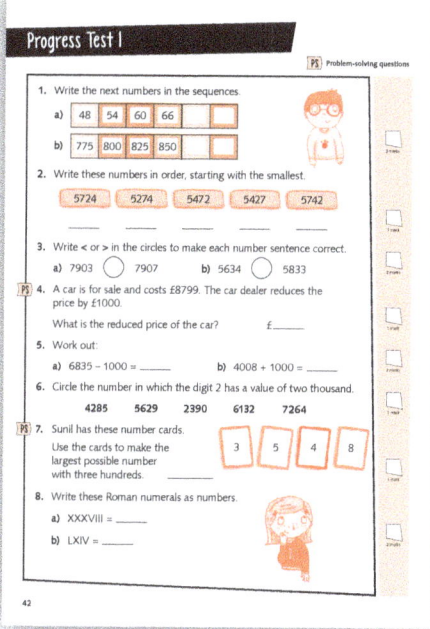
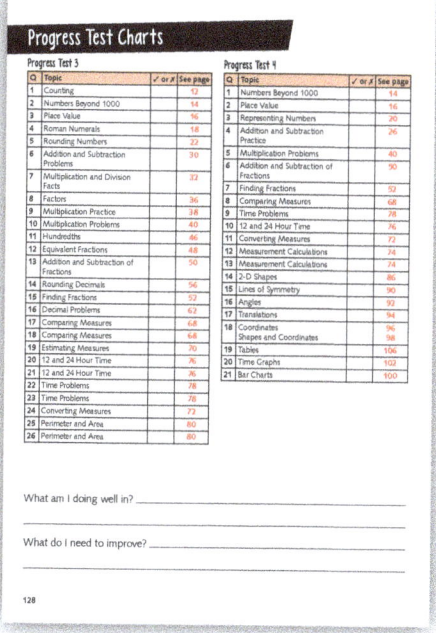

Answers for all the questions are included in a pull-out answer section at the back of the book.

Contents

Starter Test 4

Number – Number and Place Value
Counting 12
Numbers Beyond 1000 14
Place Value 16
Roman Numerals 18
Representing Numbers 20
Rounding Numbers 22
Negative Numbers 24

Number – Addition and Subtraction
Addition and Subtraction Practice 26
Estimating and Checking Calculations 28
Addition and Subtraction Problems 30

Number – Multiplication and Division
Multiplication and Division Facts 32
Mental Multiplication and Division 34
Factors 36
Multiplication Practice 38
Multiplication Problems 40

Progress Test 1 42

Fractions (including Decimals)
Hundredths 46
Equivalent Fractions 48
Addition and Subtraction of Fractions 50
Finding Fractions 52
Fraction and Decimal Equivalents 54
Rounding Decimals 56
Comparing Decimals 58
Dividing by 10 and 100 60
Decimal Problems 62

Progress Test 2 64

Starter Test

PS 8. Ola works in a park. She plants 287 bulbs on Monday and 348 on Tuesday.

How many bulbs has she planted altogether? _____

1 mark

9. Write the missing numbers.

   a) 628 – _____ = 185   b) 384 + _____ = 768

2 marks

10. Harry calculates 428 + 265 = 693

    Circle the inverse calculation Harry uses to check this calculation.

    | 265 + 428 | 693 + 265 | 693 – 265 | 428 – 265 |

1 mark

11. Work out these multiplications.

    a) 18 × 3 = _____   b) 24 × 4 = _____   c) 14 × 8 = _____

3 marks

12. Work out these divisions.

    a) 70 ÷ 5 = _____   b) 76 ÷ 4 = _____   c) 54 ÷ 3 = _____

3 marks

PS 13. Carla has 4 boxes of drink cans.
Each box holds 18 cans.

How many drink cans does she have altogether? _____

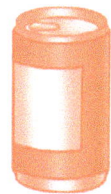

1 mark

PS 14. 69 oranges are put into packs of 3.

How many packs will there be? _____

1 mark

PS 15. 5 sweets cost 12p.

How much would 20 sweets cost? _____p

1 mark

16. Write the fractions the arrows point to.

    a) _____ b) _____

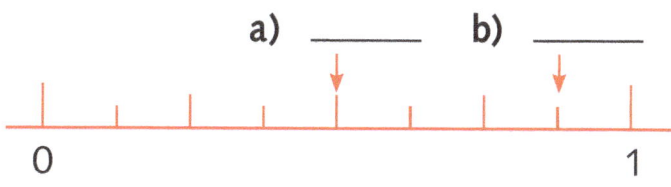

2 marks

Starter Test

PS Problem-solving questions

PS **17.** Josef cuts a cake into ten pieces.

What fraction of the whole cake is one piece? _____

1 mark

18. Jenny buys 20 apples. $\frac{1}{5}$ of the apples are red; the rest are green.

a) How many red apples are there? _____

b) How many green apples are there? _____

c) What fraction of the apples are green? _____

3 marks

19. Write the fraction each arrow points to.

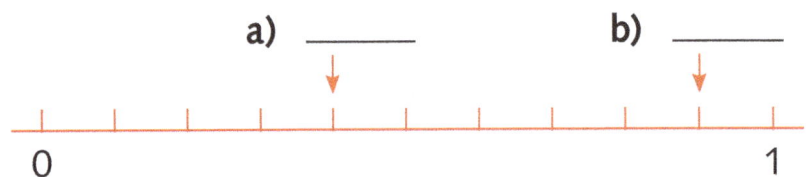

2 marks

20. Use the diagrams to find the missing numbers in these equivalent fractions.

a) [diagram] $\frac{1}{3} = \frac{\square}{6}$ b) [diagram] $\frac{1}{2} = \frac{\square}{8}$

2 marks

21. Write < or > in the circles to make each number sentence correct.

a) $\frac{1}{3}$ ◯ $\frac{2}{3}$ b) $\frac{7}{8}$ ◯ $\frac{3}{8}$

2 marks

22. Work out:

a) $\frac{1}{5} + \frac{1}{5} =$ _____   b) $\frac{9}{10} - \frac{6}{10} =$ _____   c) $\frac{1}{8} + \frac{6}{8} =$ _____

3 marks

PS **23.** A pizza is cut into ten pieces. Danny eats two of the pieces. Complete this sentence.

Danny has eaten _____-fifth of the pizza.

1 mark

PS **24.** Samir goes on a walk. He completes $\frac{3}{8}$ of the walk before lunch. He does a further $\frac{4}{8}$ of the walk before he stops for a rest.

What fraction of the walk has he completed? _____

1 mark

Starter Test

25. Write these lengths in order, starting with the smallest.

1 cm   1 m   1 mm

1 mark

26. Circle the longest length.

50 cm   100 mm   2 m   10 cm

1 mark

27. Write < or > in the circles to make each number sentence correct.

a) 4 kg ◯ 2000 g   b) 3000 g ◯ 2 kg

2 marks

(PS) 28. Kate has three parcels. She writes the weights on the side of each parcel. Tick the lightest parcel.

1½ kg   900 g   0.5 kg

1 mark

29. Write < or > in the circles to make each number sentence correct.

a) 2 litres ◯ 400 ml   b) 5000 ml ◯ 1 litre

2 marks

(PS) 30. Chris has three jugs. They hold 2000 ml, 4 litres and 1.5 litres.

Write the largest capacity. _____

1 mark

31. What is the mass of the parcel?

_____ kg

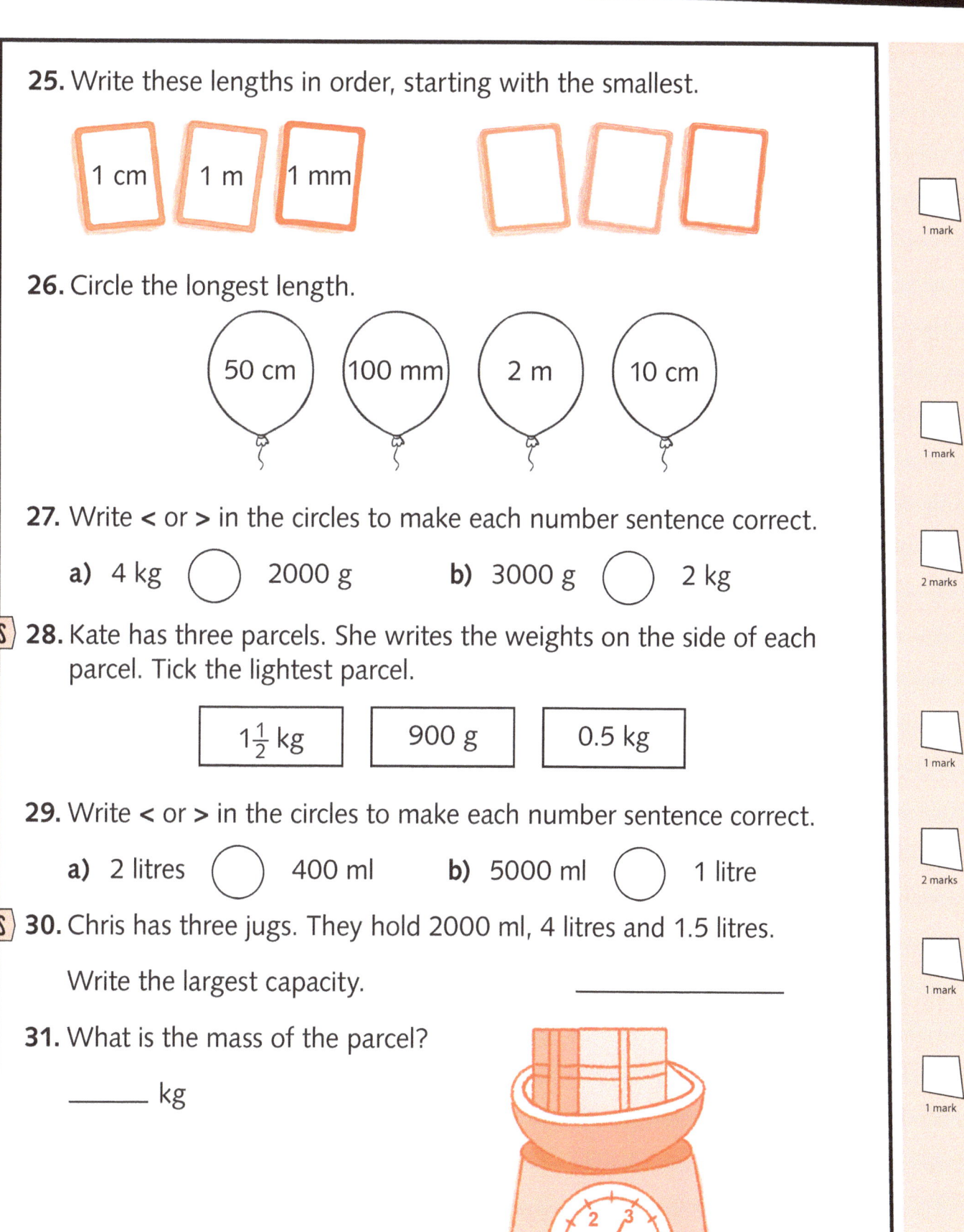

1 mark

Starter Test

PS Problem-solving questions

32. How much water is in each jug?

a) 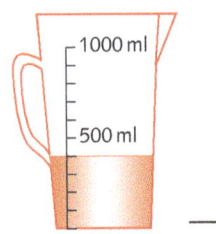 _____ ml

b) 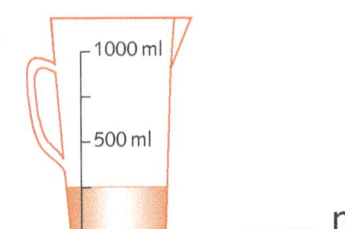 _____ ml

33. What time is shown on each clock? Write the time in 12 hour time.

a)  _____

afternoon time

b)  _____

morning time

c)  _____

morning time

d)  _____

evening time

PS 34. a) Kirstie says, "I went to bed at quarter past nine last night."

Write this time in 12 hour time. _____

b) Sophie says, "I set my alarm clock for five to seven in the morning."

Write this time in 12 hour time. _____

PS 35. Kira boils an egg for 3 minutes.

How many seconds is this? _____ seconds

36. Circle the last day in September.

| 28th September | 29th September | 30th September | 31st September |

8

Starter Test

37. 2019 is not a leap year.

How many days will there be in 2019? _____ days

PS **38.** Javid's PE lesson begins at 2:20pm and finishes at 3:10pm.

How long does the PE lesson last? _____ minutes

39. Measure the perimeter of this rectangle.

_____

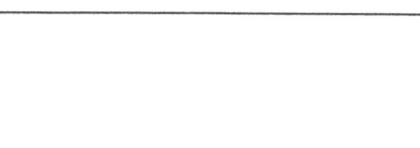

PS **40.** Yasmin has £30 and spends £17.70.

How much money does she have left?   £_____

PS **41.** Paul has £35 in his wallet, £4.85 in his pocket and he gets £50 from the bank.

How much money does he have altogether?   £_____

PS **42.** Work out:

a) 5 m + 300 cm = _____ cm   b) 7 kg + 4000 g = _____ kg

c) 6000 ml − 4 litres = _____ litres   d) 5 cm − 5 mm = _____ mm

43. Look at the red line.

Tick the line that is parallel to it.   A   B    C

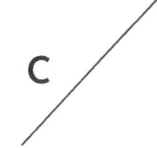

44. Each hexagon has a pair of lines drawn inside it.

Circle the pair of lines that are perpendicular to each other.

A 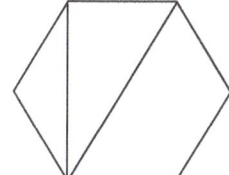   B 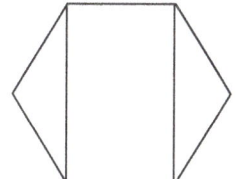   C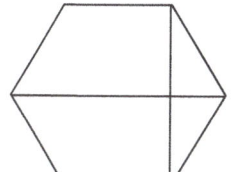

Starter Test

PS Problem-solving questions

45. Here is a squared grid.
Draw a pentagon with only one right angle.

1 mark

46. Choose the correct name from the words below for each 3-D shape. Then answer the remaining questions.

cylinder   cone   cube   cuboid

a) What is shape **A**? _____

b) What is shape **B**? _____

c) How many faces does shape **A** have? _____

d) What shape is the shaded face on shape **B**? _____

4 marks

47. Tick the angles that are greater than a right angle.

A   B   C   D

2 marks

48. How many right angles are in a half turn? _____ right angles

1 mark

PS **49.** This table shows the numbers of pieces of fruit sold in a school break.

	Apples	Pears	Bananas
Boys	23	15	28
Girls	28	9	27

a) How many bananas were sold? _____ bananas

b) How many pieces of fruit were bought by girls? _____ pieces of fruit

2 marks

10

Starter Test

50. This graph shows the number of lengths some friends swam. Altogether they swam 25 lengths.

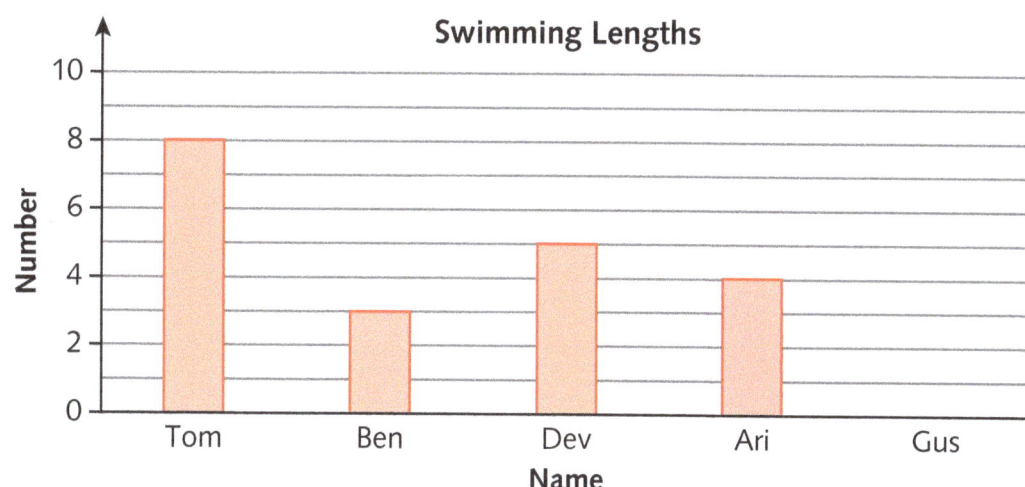

a) Who swan 3 lengths? _____

b) How many more lengths did Tom swim than Dev? _____ lengths

c) How many lengths did Gus swim? _____ lengths

d) Complete the bar chart with the bar for Gus.

4 marks

51. This pictogram shows the number of dogs that live in four streets.

	🐕 represents 2 dogs
Street	**Number of dogs**
Oak Street	🐕 🐕 🐕 🐕
Ivy Street	🐕 🐕 🐕
Elm Street	🐕 🐕 🐕 🐕 🐕
Ash Street	🐕 🐕 🐕 🐕

a) How many dogs live in Ash Street? _____ dogs

b) How many more dogs live in Elm Street than in Ivy Street? _____ dogs

c) A new street is added. 12 dogs live in Birch Street.

How many dog symbols will show 12 dogs? _____ symbols

3 marks

Marks........./89

Number – Number and Place Value

Counting

PS Problem-solving questions

Challenge 1

1 Fill in the missing terms in these sequences.

a) 0   8   16   24   32   ____   ____

b) 18   24   30   36   42   ____   ____

2 marks

2 What is the rule for each sequence of numbers?

a)  6 12 18 24 30   _____

b) 250 275 300 325 350   _____

2 marks

PS **3** Nia is counting in sevens. Will 80 be in Nia's sequence? _____

1 mark

PS **4** Poppy counts in sixes.

36   42   48   56   60

Which number is incorrect? _____

1 mark

Marks......... /6

Challenge 2

1 Fill in the missing terms in these sequences.

a) 18   24   30   36   ____   ____

b) 150   175   ____   ____   250   275

2 marks

2 What is the rule for each sequence of numbers?

a) 70   77   84   91   98   _____

b) 400   425   450   475   500   _____

2 marks

12

Number – Number and Place Value

Counting

PS **3** This number grid is damaged.

What number would be in the last square? _____

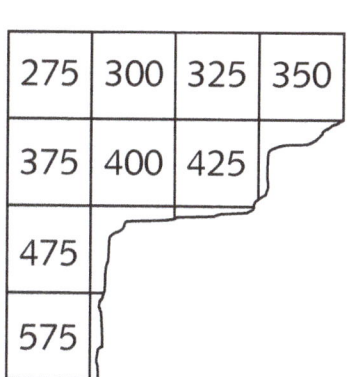

1 mark

Marks......... /5

Challenge 3

1 Fill in the next three terms in this sequence.

1 mark

2 The numbers in this sequence add the same number. Fill in the missing terms.

1 mark

PS **3** Polly is counting in nines: 45   54   63   71   81   90

Which number is incorrect? _____

1 mark

4 Here is a number grid.
Moving right the numbers add 7.
Moving down the numbers add 9.
Fill in the missing numbers.

Add 7 →

Add 9 ↓

21	28	35	42	49
30	37	44		58
39		53	60	67
48	55	62	69	76

1 mark

Marks......... /4

Total marks .... /15   How am I doing?

13

Number – Number and Place Value

Numbers Beyond 1000

PS Problem-solving questions

Challenge 1

1. Write < or > in each circle to make these statements correct.

   a) 4308 ◯ 4380   b) 3681 ◯ 3618   c) 5283 ◯ 5286

   3 marks

2. Tick the correct statement.

   A 3670 < 3669 ☐   B 5410 < 3409 ☐   C 1872 < 2890 ☐

   1 mark

3. Write these numbers in order, starting with the smallest.

   3563   4645   4536   4654   3465

   _____ _____ _____ _____ _____

   1 mark

PS 4. Gary has these four digit cards.

   5   7   4   1

   What is the largest four-digit number Gary can make? _____

   1 mark

Marks.......... /6

Challenge 2

1. Write < or > in each circle to make these statements correct.

   a) 5437 ◯ 4437   b) 2718 ◯ 2721   c) 4324 ◯ 4342

   3 marks

2. Write these numbers in order, starting with the smallest.

   4831   4138   5642   5246   5264

   _____ _____ _____ _____ _____

   1 mark

Number – Number and Place Value

Numbers Beyond 1000

PS 3 Tao has these five digit cards.

a) What is the largest three-digit number Tao can make? _____

b) What is the smallest four-digit number Tao can make? _____

2 marks

Marks .... /6

Challenge 3

1 Write these numbers in order, starting with the smallest.

    681   7861   12 612   5891   11 937

_____ _____ _____ _____ _____

1 mark

PS 2 Manisha writes four numbers in order, starting with the smallest.

    6278   6285   ?   6290

What could the missing number be? _____

1 mark

PS 3 The number of people at five football matches is recorded.

Match	Number of people
Rovers vs Town	12 834
City vs Rangers	9452
United vs Albion	10 099
County vs North End	13 901
Athletic vs Wanderers	9078

Which match had the smallest number of people?

_____

1 mark

Marks .... /3

Total marks .... /15   How am I doing?

15

Number – Number and Place Value

Place Value

PS Problem-solving questions

Challenge 1

1 What is the value of the 4 in each of these numbers?

a) 3549 _____   b) 7427 _____   c) 8194 _____

3 marks

2 Circle the number that has a digit with a value of 7 tens.

5752   8967   1475

1 mark

3 Circle the number that has a digit with a value of 3 hundreds.

36   341   4103

1 mark

PS **4** Arrange these numbers to make a four-digit number. _____

7 ones   6 hundreds   2 tens   3 thousands

1 mark

Marks.......... /6

Challenge 2

1 What is the value of the 9 in each of these numbers?

a) 7920 _____   b) 6295 _____   c) 9302 _____

3 marks

2 Circle the number that has a digit with a value of 8 hundreds.

8006   3890   1080

1 mark

3 Write the total of 8 + 700 + 30 + 9000. _____

1 mark

4 Write the total of 4 tens + 3 hundreds + 2 ones + 6 thousands. _____

1 mark

16

Number – Number and Place Value

Place Value

PS 5 Here are five digit cards.

4 3 5 6 2

a) Use any four cards to make a number larger than 5000. _____

b) Use any four cards to make the smallest four-digit number possible. _____

2 marks

Marks .... /8

Challenge 3

PS 1 Jamie says, "4021 is smaller than 897 because it has smaller numbers."

Explain to Jamie why he is incorrect.

_____

_____

1 mark

PS 2 Nisha is thinking of a four-digit number.
Use the clues below to work out Nisha's number. _____

- When you add the four digits, the total is 10.
- The thousands digit is 6.
- The tens digit is half the thousands digit.
- The ones digit is 1.

1 mark

3 Work out the number that is:

a) one ten more than 9999. _____

b) one hundred more than 9990. _____

2 marks

Marks .... /4

Total marks .... /18   How am I doing?

Number – Number and Place Value

Roman Numerals

PS Problem-solving questions

Challenge 1

1 Change these Roman numerals to numbers.

a) XII _____   b) XV _____   c) XXVI _____

d) VIII _____   e) XXXIII _____   f) XL _____

6 marks

2 Here are some Roman calculations. Write them using numbers.

a) III + II = V   b) VI + IV = X

___ + ___ = ___   ___ + ___ = ___

c) XXVI – VII = XIX

___ – ___ = ___

3 marks

PS **3** Write the time shown on each clock.

a)    b)

_____   _____

2 marks

Marks......... /11

Challenge 2

1 Change these Roman numerals to numbers.

a) LVI = _____   b) XLVIII = _____

c) XCVI = _____   d) LXXXIII = _____

4 marks

2 Write the next Roman numeral in the sequence as a number.

LXXXVI   LXXXVII   LXXXVIII   LXXXIX   _____

1 mark

Number – Number and Place Value

Roman Numerals

3 Here are some Roman calculations. Write them using numbers.

a) XL + LX = C _____ + _____ = _____

b) LXIX − XXX = XXXIX _____ − _____ = _____

2 marks

4 Circle the Roman numeral that is equivalent to 99.

IC   CX   XCXI   XCIX   LXLIX

1 mark

Marks .... /8

Challenge 3

1 Write these Roman numerals in order of size, smallest first.

XC   LXV   XXXVIII   LXIX   XLVII

smallest [ ] [ ] [ ] [ ] [ ] largest

1 mark

2 Complete these calculations. Write your answers as numbers.

a) LXIV − XLVII + XXIX = _____

b) XXIV + XIX + XCIII = _____

c) XL + LX − L = _____

d) XXXIX − XXI + LXVI = _____

4 marks

3 These Roman numerals are larger than 100. Write them as numbers.

a) CXXV _____   b) CCXXII _____

c) CCCLIX _____

3 marks

Marks .... /8

Total marks .... /27   How am I doing?

Number – Number and Place Value

Representing Numbers

PS Problem-solving questions

Challenge 1

1 Estimate the numbers the arrows point to.

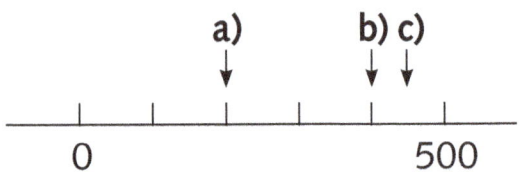

a) _____

b) _____

c) _____

3 marks

2 Write the number shown on each abacus.

a)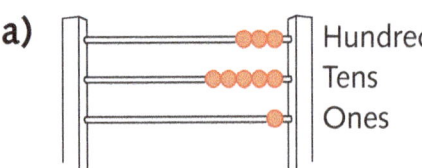

b)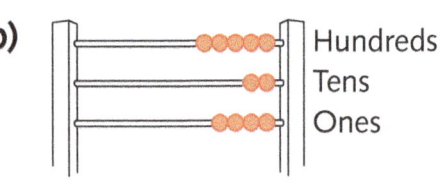

_____   _____

2 marks

3 Write these numbers using digits.

a) Five thousand, two hundred and ninety-six = _____

b) Seven thousand, five hundred and seventy = _____

c) Two thousand, eight hundred and four = _____

3 marks

Marks.......... /8

Challenge 2

1 Write the three-digit number represented by these blocks. _____

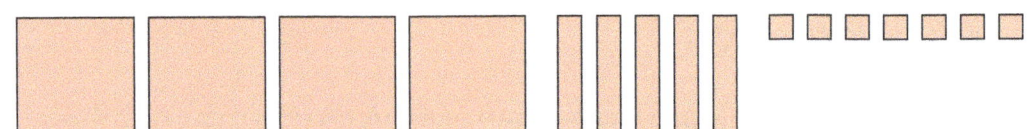

1 mark

2 Look at this calculation: 2538 = 2000 + 500 + 30 + 8

Write these numbers in a similar way.

a) 4825 = _____ + _____ + _____ + _____

b) 6719 = _____ + _____ + _____ + _____

2 marks

Number – Number and Place Value

Representing Numbers

3 Estimate the numbers the arrows point to.

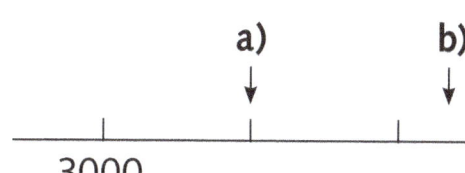

a) _____

b) _____

c) _____

3 marks

Marks .... /6

Challenge 3

PS 1 6538 can be written as 6000 + 400 + 130 + 8.

Find five ways to write 3746.

a) _____   b) _____

c) _____   d) _____

e) _____

5 marks

2 Write these numbers as digits.

a) Four thousand, six hundred and nine _____

b) Eight thousand and twenty _____

2 marks

3 Complete the following.

a) 7000 = _____ thousands = _____ hundreds

= _____ tens = _____ ones

b) 80 tens = _____   c) 30 hundreds = _____

3 marks

4 Use these place value cards to write a four-digit number. _____

| 7 hundreds | 8 ones | 2 thousands | 3 tens |

1 mark

Marks .... /11

Total marks .... /25   How am I doing?

Number – Number and Place Value

Rounding Numbers

PS Problem-solving questions

Challenge 1

1 Round each number to the nearest 10.

a) 54 _____   b) 99 _____   c) 76 _____

3 marks

2 Enrico is rounding numbers to the nearest 10.
Circle the numbers he will round to 80.

72   84   75   85   77

3 marks

PS **3** Jo estimates the answer to 52 + 74 by rounding both numbers to the nearest 10 and adding.

Write Jo's estimated answer. _____

1 mark

4 Round each number to the nearest 100.

a) 326 _____   b) 704 _____   c) 1477 _____

3 marks

Marks......... /10

Challenge 2

1 Round each number to the nearest 100.

a) 3783 _____   b) 6317 _____   c) 5078 _____

3 marks

2 Round 2628 to the:

a) nearest 10. _____

b) nearest 100. _____

c) nearest 1000. _____

3 marks

3 Round each number to the nearest 1000.

a) 4009 _____   b) 2505 _____   c) 8199 _____

3 marks

22

Number – Number and Place Value

Rounding Numbers

4 Sean rounded a number to the nearest 100. His answer was 700.

Write three different numbers that Sean's number could have been.

_____   _____   _____

3 marks

Marks........./12

Challenge 3

1 Round each number to the nearest 1000.

a) 6043 _____   b) 789 _____   c) 3911 _____

3 marks

PS 2 Max has a secret number.
He rounds the number down to the nearest 1000, the answer is 7000.
He rounds the same number to the nearest 10, the answer is 7340.

Write a possible secret number. _____

1 mark

3 Oliver rounds numbers to the nearest 10.
Circle the numbers Oliver has rounded.

350   4005   6000   7900   10 101

3 marks

4 Eve has four numbers cards.
She makes four-digit numbers using the cards and rounds them to the nearest 1000.

5   9   6   1

Write the 4 four-digit numbers that Eve will round to 7000.

_____ _____ _____ _____

4 marks

Marks........./11

Total marks ..../33   How am I doing?

23

Number – Number and Place Value

Negative Numbers

PS Problem-solving questions

Challenge 1

```
-6  -5  -4  -3  -2  -1   0   1   2   3   4   5   6
 |   |   |   |   |   |   |   |   |   |   |   |   |
```

1 Use the number line.

   a) Start at 3, count back 5, write the number you reach. _____

   b) Start at 2, count back 6, write the number you reach. _____

   c) Start at 5, count back 7, write the number you reach. _____

   d) Start at 4, count back 9, write the number you reach. _____

   4 marks

2 Calculate by counting backwards:

   a) 4 – 6 = _____   b) 3 – 7 = _____   c) 5 – 6 = _____

   3 marks

PS 3 Jake wants to buy a t-shirt for £7. He has £3.
He borrows the rest from his dad.
How much does Jake owe his dad? _____

   1 mark

4 Start at –2, count back 3, write the number you reach. _____

   1 mark

Marks.......... /9

Challenge 2

1 a) Start at 4, count back 8, write the number you reach. _____

   b) Start at 2, count back 9, write the number you reach. _____

   c) Start at 10, count back 12, write the number you reach. _____

   d) Start at 1, count back 10, write the number you reach. _____

   4 marks

2 Calculate by counting backwards:

   a) 2 – 7 = _____   b) 12 – 17 = _____   c) 15 – 16 = _____

   3 marks

3 Calculate by counting backwards:

   a) –2 – 7 = _____   b) –5 – 12 = _____   c) –10 – 8 = _____

   3 marks

24

Number – Number and Place Value

Negative Numbers

PS **4** Look at the thermometer.

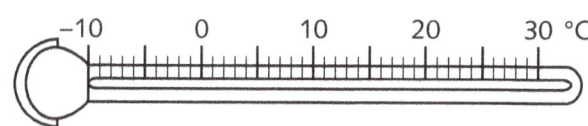

Write the temperature after these changes.

a) The temperature is 12°C and falls by 14°C. _____

b) The temperature is −4°C and falls by 5°C. _____

c) The temperature is −3°C and rises by 7°C. _____

3 marks

Marks.......... /13

Challenge 3

1 Arrange these numbers in order, starting with the largest.

−6   5   −1   −4   3

largest [ ][ ][ ][ ][ ] smallest

1 mark

2 This number line runs from −10 to 10. Four arrows point to whole numbers. Write the numbers the arrows point to.

a) _____   b) _____   c) _____   d) _____

4 marks

3 Calculate:

a) 3 − 12 = _____   b) 5 − 20 = _____   c) 12 − 20 = _____

d) −5 − 9 = _____   e) −12 − 12 = _____   f) −8 + 4 = _____

6 marks

Marks.......... /11

Total marks .... /33   How am I doing?

Number – Addition and Subtraction

Addition and Subtraction Practice

Challenge 1

1 Work out the answers to these addition problems.

a)   3 2 7
   + 4 2 2
   ———

b)   7 9 2
   + 2 0 5
   ———

c)   4 9 2
   + 5 1 3
   ———

2 Work out the answers to these subtraction problems.

a)   6 5 8
   − 2 5 7
   ———

b)   5 8 8
   − 4 9 7
   ———

c)   7 0 1
   − 1 7 3
   ———

3 marks

3 marks

Marks......... /6

Challenge 2

1 Work out the answers to these addition problems.

a)   8 5 4
   + 6 7 4
   ———

b)   8 3 9
   + 5 7 9
   ———

c)   8 8 4 4
   + 4 8 8 4
   ———

3 marks

26

Addition and Subtraction Practice

Number – Addition and Subtraction

2 Work out the answers to these subtraction problems.

a) 9 0 3
   − 2 6 0
   ─────────

b) 7 9 0
   − 3 8 4
   ─────────

c) 5 3 3 1
   − 1 7 3 5
   ─────────

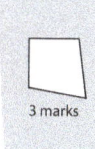
3 marks

Marks.......... /6

Challenge 3

1 Work out the answers to these addition problems.

a) 5 9 2 1
   + 6 4 7 9
   ─────────

b) 8 5 3 6
   + 2 7 7 6
   ─────────

c) 6 4 3 7
   + 6 4 7 2
   ─────────

3 marks

2 Work out the answers to these subtraction problems.

a) 6 6 7 3
   − 5 4 7 7
   ─────────

b) 9 0 6 3
   − 5 0 9 0
   ─────────

c) 8 4 7 2
   − 3 6 7 9
   ─────────

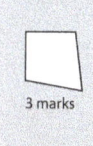
3 marks

Marks.......... /6

Total marks .... /18   How am I doing?

Number – Addition and Subtraction

Estimating and Checking Calculations

PS Problem-solving questions

Challenge 1

1 Round each number to the nearest 10. Then use the rounded numbers to give an estimated answer to each calculation.

  a) 54 + 74 _____

  b) 33 + 64 _____

  c) 76 + 85 _____

3 marks

PS 2 Tara calculates 125 + 265 = 390. She wants to check her calculation by calculating the inverse.

Circle the calculation that is the inverse of Tara's calculation.

| 265 + 390 | 390 – 265 | 390 + 125 | 265 – 125 |

1 mark

3 Write the inverse of each calculation.

  a) 235 + 328 = 563 b) 529 – 236 = 293 c) 525 – 214 = 311

  _____   _____   _____

3 marks

PS 4 A gardener plants tulip and daffodil bulbs. Altogether, there are 486 bulbs. She plants 259 tulip bulbs.

How many are daffodil bulbs? _____

1 mark

Marks.......... /8

Challenge 2

PS 1 Joshua thinks of a number. He adds 56 to the number. Joshua's answer is 412.

What was Joshua's number? _____

1 mark

PS 2 Kia thinks of a number. She subtracts 94 from the number. Kia's answer is 536.

What was Kia's number? _____

1 mark

28

Number – Addition and Subtraction
Estimating and Checking Calculations

3 Estimate the answers to these calculations by rounding to the nearest 100.

a) 418 + 629 _____ b) 907 – 798 _____

2 marks

4 Estimate the answers to these calculations by rounding to the nearest 1000.

a) 6321 – 2976 _____ b) 6602 + 3532 _____

2 marks

Marks.......... /6

Challenge 3

PS 1 Bryn estimates the price of two pens as 50p and 80p by rounding to the nearest 10p. The actual cost of the two pens is £1.35.

What could the price of each pen have been? _____ and _____

1 mark

PS 2 Estimate answers to these calculations by rounding to the nearest 1000.

a) 7286 + 9781 _____ b) 12 084 – 6822 _____

2 marks

PS 3 Ari completes this calculation: 6432 + 3826 = 9258

Use an inverse calculation to check Ari's calculation. If it's wrong, correct the answer.

_____

1 mark

PS 4 Dino adds a number and 732, then subtracts 218 and reaches an answer of 861.

Write the number that Dino started with. _____

1 mark

Marks.......... /5

Total marks .... /19   How am I doing?

Number – Addition and Subtraction

Addition and Subtraction Problems

PS Problem-solving questions

Challenge 1

PS 1 In a school, there are 286 children in Key Stage 2 and 139 children in Key Stage 1.

a) How many more children are there in Key Stage 2 than in Key Stage 1?

_____ children

b) How many children are there in Key Stages 1 and 2 altogether?

_____ children

2 marks

PS 2 A town has two car parks. One car park holds 327 cars; the other holds 286 cars.

How many parking spaces are there altogether? _____ spaces

1 mark

PS 3 Nisha chooses between two digital devices. One costs £439 and the other costs £385.

What is the difference in price between the two devices?   £ _____

1 mark

PS 4 Nino is 161 cm tall. His brother is 36 cm shorter.

How tall is Nino's brother?   _____ cm

1 mark

Marks.......... /5

Challenge 2

PS 1 Work out the number that is 3965 greater than 2832. _____

1 mark

PS 2 A theatre group gave two shows. In the afternoon, there were 1568 people in the audience and in the evening there were 1735 people.

How many people watched the two performances? _____ people

1 mark

30

Number – Addition and Subtraction

Addition and Subtraction Problems

 3 Noah went on holiday. He drove 864 km from London to the south of France. He drove 365 km whilst on holiday and then he drove back to London.

How many kilometres did he drive altogether? _____ km

1 mark

 4 A shop has a stock of 3475 t-shirts. It sells 1294 t-shirts.

How many t-shirts does it have left? _____ t-shirts

1 mark

Marks........../4

Challenge 3

 1 Write the missing numbers in these calculations.

a)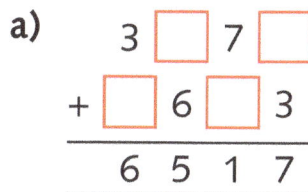

b)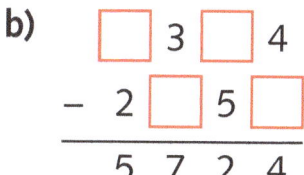

2 marks

 2 Here are six number cards:

Use the cards to complete this calculation.

 = 840

1 mark

 3 Work out these calculations and circle the one with the greatest answer.

A | 5236 + 1609 = _____

B | 9343 − 2434 = _____

C | 2354 + 4567 = _____

4 marks

Marks........../7

Total marks ..../16   How am I doing?

31

Number – Multiplication and Division

Multiplication and Division Facts

PS Problem-solving questions

Challenge 1

1 Write the answers to:

a) 7 × 5 = _____   b) 3 × 8 = _____

c) 11 × 8 = _____   d) 12 × 3 = _____

e) 7 × 10 = _____   f) 8 × 4 = _____

6 marks

2 Fill in the missing numbers.

a) 4 × _____ = 28   b) 5 × _____ = 60

c) 3 × _____ = 21   d) _____ × 8 = 48

e) _____ × 10 = 100   f) _____ × 4 = 24

6 marks

3 Write the answers to:

a) 16 ÷ 2 = _____   b) 45 ÷ 5 = _____

c) 36 ÷ 4 = _____   d) 56 ÷ 8 = _____

e) 18 ÷ 3 = _____   f) 48 ÷ 4 = _____

6 marks

Marks......... /18

Challenge 2

1 Write the answers to:

a) 7 × 6 = _____   b) 4 × 7 = _____

c) 9 × 11 = _____   d) 8 × 7 = _____

e) 9 × 9 = _____   f) 12 × 7 = _____

6 marks

2 Fill in the missing numbers.

a) 7 × _____ = 49   b) 9 × _____ = 72

c) 6 × _____ = 72   d) _____ × 9 = 108

e) _____ × 12 = 144   f) _____ × 7 = 63

6 marks

32

Number – Multiplication and Division

Multiplication and Division Facts

3 Write the answers to:

a) 121 ÷ 11 = _____ b) 27 ÷ 9 = _____

c) 66 ÷ 6 = _____ d) 96 ÷ 8 = _____

e) 54 ÷ 9 = _____ f) 36 ÷ 9 = _____

6 marks

Marks......... /18

Challenge 3

1 Complete the multiplication squares.

a)

×	7	9	12
7	49		
9			
12			

b)

×		8	
6	42		
		64	
	84		108

7 marks

2 Use different whole numbers only to complete these multiplication facts.

a) 6 × 8 = _____ × _____ b) 4 × 9 = _____ × _____

c) 3 × 8 = _____ × _____ d) 6 × 12 = _____ × _____

4 marks

3 Use whole numbers only to complete these division facts.

a) 54 ÷ 6 = _____ ÷ 9 b) 64 ÷ 8 = _____ ÷ 7

c) 96 ÷ 12 = _____ ÷ 8 d) 132 ÷ 11 = _____ ÷ 10

4 marks

Marks......... /15

Total marks .... /51   How am I doing?

Number – Multiplication and Division

Mental Multiplication and Division

Challenge 1

1 Write the answers to:

a) 5 × 1 = _____ b) 10 × 0 = _____ c) 0 × 8 = _____

d) 20 × 3 = _____ e) 40 × 4 = _____ f) 70 × 5 = _____

6 marks

2 Fill in the missing numbers.

a) _____ × 5 = 100 b) _____ × 4 = 80

c) _____ × 3 = 90

3 marks

3 Write the answers to:

a) 80 ÷ 2 = _____ b) 150 ÷ 5 = _____

c) 200 ÷ 4 = _____

3 marks

4 Fill in the missing numbers.

a) 180 ÷ _____ = 60

b) 250 ÷ _____ = 50

c) 160 ÷ _____ = 40

3 marks

Marks......... /15

Challenge 2

1 Write the answers to:

a) 15 × 2 = _____ b) 22 × 5 = _____ c) 16 × 4 = _____

d) 23 × 3 = _____ e) 32 × 4 = _____ f) 51 × 5 = _____

6 marks

2 Write the answers to:

a) 45 ÷ 3 = _____ b) 52 ÷ 4 = _____ c) 70 ÷ 5 = _____

d) 68 ÷ 2 = _____ e) 64 ÷ 4 = _____ f) 66 ÷ 3 = _____

6 marks

Number – Multiplication and Division

Mental Multiplication and Division

3 Write the answers to:

a) 4 × 2 × 6 = _____

b) 5 × 2 × 9 = _____

c) 6 × 2 × 6 = _____

3 marks

Marks......... /15

Challenge 3

1 Fill in the missing numbers.

a) 23 × _____ = 69

b) 32 × _____ = 128

c) 51 × _____ = 255

d) 53 × _____ = 159

e) 24 × _____ = 120

f) 82 × _____ = 246

6 marks

2 Fill in the missing numbers.

a) 3 × 5 × _____ = 45

b) 4 × _____ × 5 = 120

c) _____ × 10 × 2 = 80

d) 7 × 3 × _____ = 84

e) 2 × _____ × 2 = 48

f) _____ × 3 × 8 = 72

6 marks

3 Fill in the missing numbers.

a) 12 × 6 = 6 × 3 × _____

b) 15 × 6 = 6 × 5 × _____

c) _____ × 8 = 8 × 5 × 3

d) _____ × 7 = 7 × 8 × 3

4 marks

Marks......... /16

Total marks .... /46   How am I doing?

35

Number – Multiplication and Division

Factors

PS Problem-solving questions

Challenge 1

1 Circle the true statements.

A 6 is a factor of 12   B 4 is a factor of 18

C 15 is a factor of 5   D 8 is a factor of 24

E 3 is a factor of 35   F 9 is a factor of 72

3 marks

2 Write the missing numbers.

a) 5 × 7 = 7 × _____   b) 12 × 4 = 4 × _____

2 marks

3 Write the missing factor of 36.

4   36   6   2   18   3   12   1   _____

1 mark

PS 4 Circle the factors of 30.

1   3   4   5   8   10   12   15

1 mark

Marks........../7

Challenge 2

1 Write all the factors of:

a) 10 _____   b) 21 _____

c) 32 _____   d) 40 _____

4 marks

2 Write the missing factor pair of 42.

| 1 and 42 | 3 and 14 | 6 and 7 | _____ and _____ |

1 mark

3 Write the missing factor pair of 56.

| 1 and 56 | 2 and 28 | 4 and 14 | _____ and _____ |

1 mark

36

Number – Multiplication and Division

Factors

4. Circle any correct statements.

   A $8 ÷ 9 = 9 ÷ 8$ B $12 × 11 = 11 × 12$

   1 mark

PS 5. I think of a number greater than 6. It is a factor of both 28 and 49.

   What is my number? _____

   1 mark

   Marks......... /8

Challenge 3

1. Write these numbers in the Venn diagram.

   4   5   6   8   10

   Factors of 20   Factors of 30

   5 marks

PS 2. Faisal says, "I have a crate of bottles. There are 8 rows of 6 bottles."

   Faisal can also say, "I have a crate of bottles. There are _____

   rows of _____ bottles."

   1 mark

3. List the factor pairs of:

   a) 48 _____

   b) 60 _____

   c) 100 _____

   3 marks

4. Write the factors of 56 that are not factors of 28. _____

   1 mark

   Marks......... /10

Total marks .... /25   How am I doing?

Number – Multiplication and Division

Multiplication Practice

PS Problem-solving questions

Challenge 1

1 Calculate these multiplications.

a) 4 6
×   5
___

b) 5 3
×   6
___

c) 6 2
×   4
___

d) 3 7
×   3
___

4 marks

2 Set out these problems using column multiplication and calculate.

a) 36 × 6 =	b) 72 × 5 =	c) 84 × 8 =

3 marks

3 Write in the missing numbers.

a) 57 × 4 = 50 × 4 + 7 × _____ = 228

b) 73 × _____ = 70 × 3 + 3 × 3 = 219

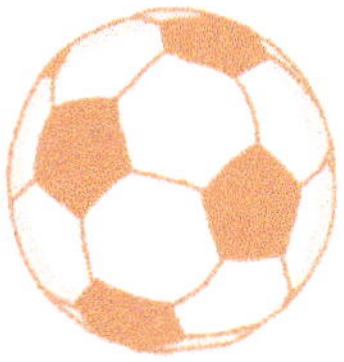

2 marks

Marks………/9

Challenge 2

1 Calculate these multiplications.

a) 2 5 3
×   6
___

b) 6 0 7
×   5
___

c) 6 3 7
×   7
___

3 marks

2 Write in the missing numbers.

a) 725 × 6 = 700 × _____ + 20 × _____ + 5 × _____ = 4350

b) 321 × 4 = 642 × _____

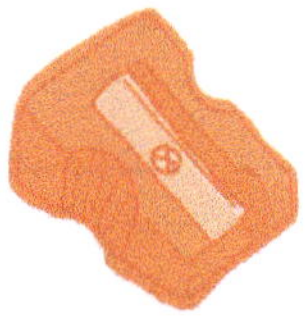

2 marks

Multiplication Practice

Number – Multiplication and Division

3 Set out these problems using column multiplication and calculate.

| a) 427 × 7 = | b) 683 × 6 = | c) 427 × 9 = |

3 marks

Marks .... /8

Challenge 3

1 Set out these problems using column multiplication and calculate.

| a) 856 × 7 = | b) 967 × 8 = | c) 956 × 9 = |

3 marks

PS 2 Fill in the missing numbers.

a) ☐ 4 ☐
    ×   7
    ─────────
    3 8 2 9

b) ☐ 8 ☐
    ×   9
    ─────────
    3 4 6 5

c) ☐ 7 ☐
    ×   8
    ─────────
    2 2 0 8

3 marks

PS 3 Nishi has four number cards. She arranges them into a multiplication.

3   4   6   6

Fill in the multiplication on the right to find the largest total Nishi can make.

☐☐☐
    ☐
× ☐☐☐☐

1 mark

Marks .... /7

Total marks .... /24   How am I doing? 😊 😐 😣

Number – Multiplication and Division

Multiplication Problems

PS Problem-solving questions

Challenge 1

PS 1 There are 48 bottles in a crate.

How many bottles will there be in 6 crates? _____ bottles

PS 2 Jack has 35 marbles in a bag. He has 8 similar bags.

How many marbles does he have altogether? _____ marbles

PS 3 Jess buys 3 packs of apples. There are 15 apples altogether.

How many apples will there be in 12 packs? _____ apples

PS 4 A pack of 4 tins of beans costs £1.75.

How much would you pay for 20 tins of beans? £_____

PS 5 There are 6 classes in a school. Each class has 28 children.

How many children are in the school? _____ children

Marks.......... /5

Challenge 2

PS 1 A farmer keeps 186 chickens. Each chicken lays 4 eggs a week.

How many eggs are laid in a week? _____ eggs

PS 2 A sack of tulip bulbs holds 256 bulbs. Dave plants 6 sacks of bulbs.

How many bulbs has Dave planted? _____ bulbs

PS 3 A shop sells 3 televisions that cost £375 each and 7 televisions that cost £435 each.

What is the total cost of the televisions? £_____

Number – Multiplication and Division

Multiplication Problems

PS 4 A school puts on a show in the school hall. It sells 278 tickets for each performance. There are 5 performances.

How many tickets are sold? _____ tickets

PS 5 A teacher buys 8 jars of counters. Each jar holds 475 counters.

How many counters are there altogether? _____ counters

Marks........./5

Challenge 3

PS 1 This table shows the cost of different air fares from London.

To From	New York	Los Angeles	Dubai	New Delhi	Sydney
London (One way)	£346	£465	£486	£518	£648
London (Return)	£452	£573	£549	£647	£732

What is the cost of:

a) 5 one-way tickets to Dubai? £_____

b) 4 return tickets to Sydney? £_____

c) 6 one-way tickets to Los Angeles? £_____

d) 8 return tickets to New Delhi? £_____

e) 7 one-way tickets to New York? £_____

Marks........./5

Total marks ..../15   How am I doing?

Progress Test 1

PS Problem-solving questions

1. Write the next numbers in the sequences.

   a)

   b)

2. Write these numbers in order, starting with the smallest.

   5724   5274   5472   5427   5742

   ____   ____   ____   ____   ____

3. Write < or > in the circles to make each number sentence correct.

   a) 7903 ◯ 7907   b) 5634 ◯ 5833

PS 4. A car is for sale and costs £8799. The car dealer reduces the price by £1000.

   What is the reduced price of the car?   £_____

5. Work out:

   a) 6835 − 1000 = _____   b) 4008 + 1000 = _____

6. Circle the number in which the digit 2 has a value of two thousand.

   4285   5629   2390   6132   7264

PS 7. Sunil has these number cards.

   Use the cards to make the largest possible number with three hundreds. _____

8. Write these Roman numerals as numbers.

   a) XXXVIII = _____

   b) LXIV = _____

Progress Test 1

9. These Roman numerals are written as a calculation: XXXV + VII = ?
   Circle the correct answer.

           LXII   XLIIIV   LII   XLII

   1 mark

10. Write the three-digit number represented by these blocks. _____

    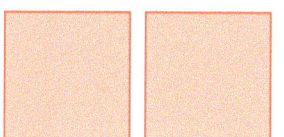

    1 mark

11. Write the numbers the arrows point to.

    a) _____   b) _____

    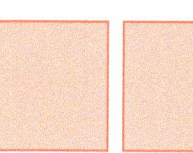

    2 marks

12. Round each number to the nearest 100.

    a) 528 _____   b) 761 _____   c) 89 _____

    3 marks

13. Complete this table by rounding the number to the nearest 10, 100 and 1000.

	Rounded to the nearest 10 is	Rounded to the nearest 100 is	Rounded to the nearest 1000 is
5826			

    3 marks

14. Write the next two numbers in this sequence.

    1 mark

PS 15. Flo counts back in fives from 20.

   Write the first number below zero she counts. _____

   1 mark

16. Work out:

    a)   4 2 7   b)   7 9 5   c)   8 1 5
          – 1 8 7   + 2 8 5   – 2 7 0

    3 marks

43

Progress Test 1

PS Problem-solving questions

PS **17.** Write in the missing numbers.

a)
$$\begin{array}{r} 5\,\square\,7 \\ -\ \square\,5\,\square \\ \hline 1\,7\,4 \\ \hline \end{array}$$

b)
$$\begin{array}{r} \square\,5\,\square \\ +\ 5\,\square\,1 \\ \hline 9\,0\,9 \\ \hline \end{array}$$

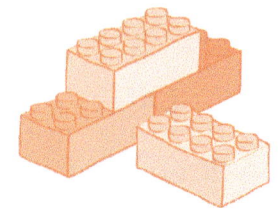

2 marks

18. Circle the calculation that is the inverse of 694 – 322 = 372

372 + 694	694 – 322	372 + 322	372 – 322

1 mark

PS **19.** Sven completes the calculation shown on the right. He then works out the inverse calculation to check his work.

$$\begin{array}{r} 3\ 8\ 2 \\ +\ 4\ 6\ 5 \\ \hline 7\ 4\ 7 \\ \hline \end{array}$$

Write the answer to Sven's inverse calculation. ______

1 mark

PS **20.** Ivete has 275 minutes talk time left on her phone. She uses up another 126 minutes and then her monthly allowance increases the talk time by 450 minutes.

How many minutes does Ivete have on her phone now? ______ minutes

1 mark

PS **21.** Tarun has to drive 356 kilometres. After 186 kilometres he stops to have lunch and then, after another 137 kilometres, he stops for a coffee.

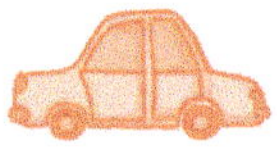

How much further does Tarun have to drive? ______ kilometres

1 mark

22. Work out:

a) 5 × 7 = ______ b) 3 × 9 = ______ c) 7 × 12 = ______

d) 48 ÷ 6 = ______ e) 144 ÷ 12 = ______ f) 54 ÷ 9 = ______

6 marks

PS **23.** Write the missing numbers in this multiplication square.

×	6	9	
4	24	36	48
		81	
7	42	63	84

2 marks

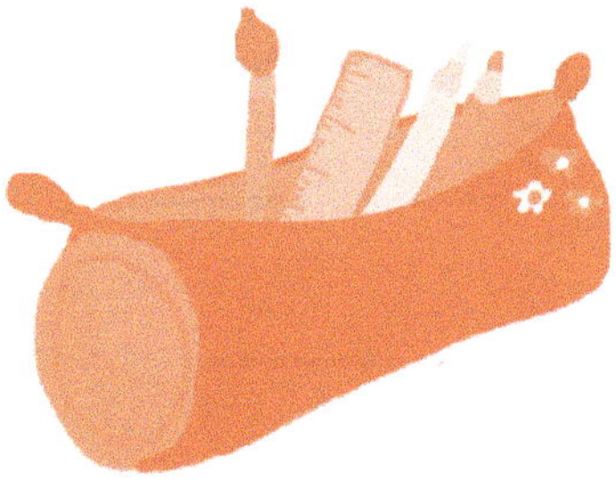

Progress Test 1

24. Work out:

    a) 15 × 5 = _____   **b)** 48 ÷ 3 = _____

    c) 22 × 4 = _____   **d)** 72 ÷ 4 = _____

    e) 96 ÷ 8 = _____   **f)** 17 × 6 = _____

6 marks

25. Write the factor pairs of:

    a) 20 _____

    b) 100 _____

2 marks

26. Work out:

    a)   6 3   **b)**   2 6 5   **c)**   6 0 8
          ×   6   ×   4   ×   7

3 marks

PS 27. A school has 8 classes and each class can hold 30 children. There are 216 children in the school.

How many spare places are there? _____ spare places

1 mark

28. Fill in the missing numbers.

54 × 3 = 50 × 3 + _____ × _____ = 162

1 mark

PS 29. A factory uses 7 metres of wood to make a wardrobe.

How much wood is needed to make 5 wardrobes? _____ metres

1 mark

30. Michael completes 3 tests worth a total of 50 marks.

How many marks are available if he completes 6 tests?

_____ marks

1 mark

Marks .... /56

Fractions (including Decimals)

Hundredths

PS Problem-solving questions

Challenge 1

1 What fraction is shown by each drawing?

a) _____   b) _____   c) _____

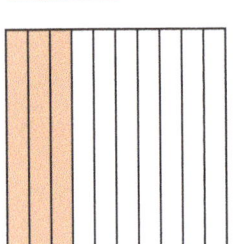 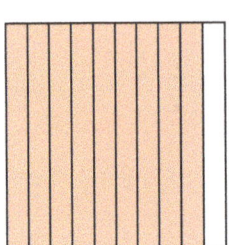 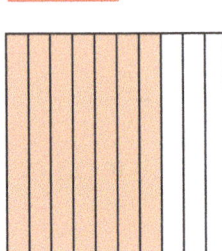

3 marks

2 Write in the missing fractions.

a)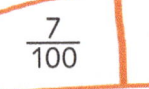

b)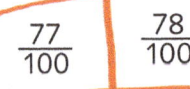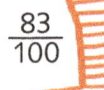

2 marks

3 Complete this sentence:

When 1 is divided by _____ the answer is $\frac{1}{100}$.

1 mark

Marks......... /6

Challenge 2

1 What fraction is shown by each drawing?

a) _____   b) _____   c) _____

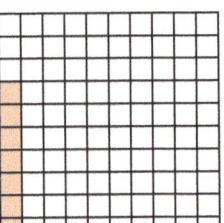 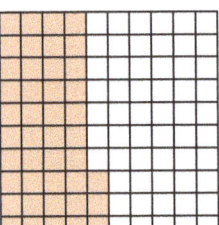 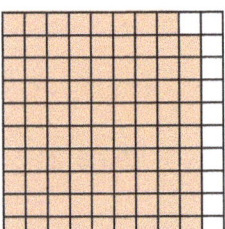

3 marks

2 Divide each number by 100. Give your answers as fractions.

a) 6 _____   b) 9 _____   c) 2 _____

3 marks

46

Fractions (including Decimals)

Hundredths

PS 3 Kesha has a 5 litre bottle of juice. This is exactly enough juice to make 100 drinks.

How much juice is used in each drink? Give your answer as a fraction of a litre.

_____

1 mark

PS 4 How many centimetres are in $\frac{1}{100}$ of a metre? _____ cm

1 mark

Marks.......... /8

Challenge 3

1 Circle the calculation with the correct answer.

$50 \div 100 = \frac{5}{100}$ $5 \div 100 = \frac{5}{10}$

$500 \div 100 = \frac{5}{100}$ $5 \div 100 = \frac{5}{100}$

1 mark

2 Circle both calculations that have an answer of $\frac{1}{100}$.

$1 \div 10$ $1 \div 100$ $0.1 \div 10$ $0.1 \div 100$

2 marks

3 Write in the missing numbers.

a) _____ $\div 100 = \frac{7}{100}$ b) _____ $\div 100 = \frac{5}{100}$

c) _____ $\div 10 = \frac{7}{100}$

3 marks

4 Write in the missing numbers.

a) $0.6 \div 10 =$ _____ $\div 100$ b) $0.7 \div$ _____ $= 7 \div 100$

2 marks

Marks.......... /8

Total marks .... /22   How am I doing?

Fractions (including Decimals)

Equivalent Fractions

 PS Problem-solving questions

Challenge 1

1 Fill in the missing equivalent fractions shown by the drawings.

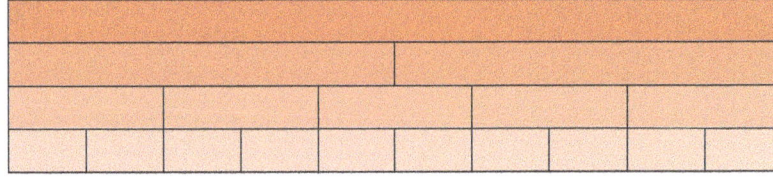

a) ____ = ____   b) ____ = ____

2 marks

2 Use the fraction bars to complete the equivalent fractions.

one whole
halves
fifths
tenths

a) $\frac{1}{2} = \frac{\square}{10}$ b) $\frac{3}{5} = \frac{\square}{10}$ c) $\frac{4}{5} = \frac{\square}{10}$

3 marks

3 Use the diagrams to help you complete the equivalent fractions.

a) 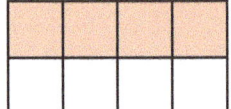   b) 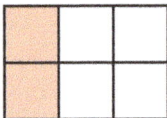   c)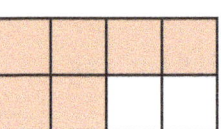

$\frac{1}{2} = \frac{4}{\square}$ $\frac{1}{3} = \frac{2}{\square}$ $\frac{\square}{4} = \frac{6}{8}$

3 marks

Marks........../8

Challenge 2

1 This rectangle is divided into twentieths.

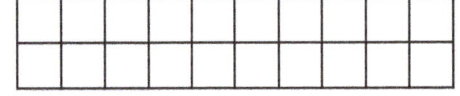

Use the diagram to help you complete the equivalent fractions.

a) $\frac{1}{2} = \frac{\square}{20}$ b) $\frac{3}{10} = \frac{\square}{20}$ c) $\frac{3}{5} = \frac{\square}{20}$

3 marks

Fractions (including Decimals)
Equivalent Fractions

2 This rectangle is divided into fifteenths. Use the diagram to help you complete the equivalent fractions.

a) $\frac{1}{3} = \frac{\Box}{15}$ b) $\frac{2}{3} = \frac{\Box}{15}$ c) $\frac{3}{5} = \frac{\Box}{15}$

3 marks

3 Here are families of equivalent fractions. Complete the fractions.

a) $\frac{1}{3} = \frac{2}{6} = \frac{3}{9} = \frac{\Box}{12} = \frac{5}{\Box} = \frac{\Box}{18} = \frac{7}{\Box} = \frac{\Box}{24}$

b) $\frac{3}{4} = \frac{6}{8} = \frac{9}{12} = \frac{\Box}{16} = \frac{15}{\Box} = \frac{\Box}{24} = \frac{21}{\Box} = \frac{\Box}{32}$

2 marks

Marks......... /8

Challenge 3

1 Use the fraction bars to help you complete the equivalent fractions.

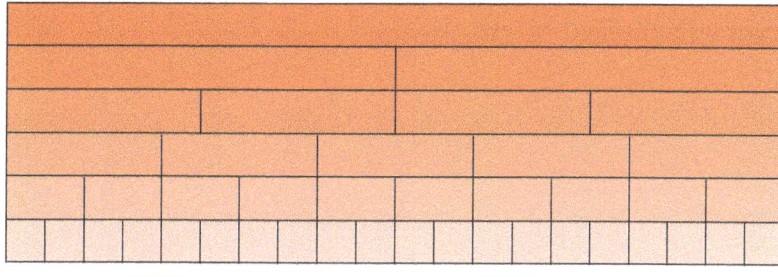

one whole / halves / quarters / fifths / tenths / twentieths

a) $\frac{1}{2} = \frac{\Box}{20}$ b) $\frac{4}{5} = \frac{\Box}{20}$ c) $\frac{3}{10} = \frac{\Box}{20}$

d) $\frac{\Box}{4} = \frac{5}{20}$ e) $\frac{\Box}{5} = \frac{8}{20}$ f) $\frac{\Box}{10} = \frac{18}{20}$

6 marks

2 Circle both fractions that are **not** equivalent to $\frac{1}{4}$.

$\frac{3}{12}$ $\frac{5}{20}$ $\frac{8}{30}$ $\frac{12}{48}$ $\frac{15}{60}$ $\frac{20}{100}$

2 marks

PS 3 There are 10 cans of paint. 4 of the cans are blue paint.

Complete this sentence: $\frac{2}{\Box}$ of the cans of paint are blue.

1 mark

Marks......... /9

Total marks .... /25   How am I doing?

Fractions (including Decimals)

Addition and Subtraction of Fractions

Challenge 1

1 Complete these calculations.

Example: ▭ + ▭ =
$\frac{1}{4}$ + $\frac{2}{4}$ = $\frac{3}{4}$

a) ▭ + ▭ =

_____ + _____ = _____

b) ▭ − ▭ =

_____ − _____ = _____

2 marks

2 Work out the answers to these calculations.

a) $\frac{2}{3} + \frac{1}{3} =$ _____   b) $\frac{7}{8} + \frac{6}{8} =$ _____   c) $\frac{8}{10} + \frac{9}{10} =$ _____

3 marks

3 Work out the answers to these calculations.

a) $\frac{4}{5} - \frac{1}{5} =$ _____   b) $\frac{9}{10} - \frac{6}{10} =$ _____   c) $\frac{7}{8} - \frac{6}{8} =$ _____

3 marks

Marks………/8

Challenge 2

1 Complete these calculations.

Example: ▭ + ▭ =
$\frac{3}{4}$ + $\frac{2}{4}$ = $\frac{5}{4}$ = $1\frac{1}{4}$

a) ▭ + ▭ =

_____ + _____ = _____ = _____

b) ▭ + ▭ =

_____ + _____ = _____ = _____

2 marks

50

Fractions (including Decimals)
Addition and Subtraction of Fractions

2 Work out the answers to these calculations.

a) $\frac{2}{3} + \frac{2}{3} =$ _____ = _____   b) $\frac{3}{5} + \frac{4}{5} =$ _____ = _____

c) $\frac{3}{4} + \frac{3}{4} =$ _____ = _____

3 marks

3 Work out the answers to these calculations.

a) $\frac{9}{10} - \frac{6}{10} =$ _____   b) $\frac{9}{12} - \frac{4}{12} =$ _____

c) $\frac{15}{10} - \frac{6}{10} =$ _____

3 marks

Marks......... /8

Challenge 3

1 Write in the missing fractions.

a)

_____ − _____ = _____

b)

_____ + _____ = _____ = _____

2 marks

2 Write the missing fractions in these calculations.

a) $\frac{2}{10} +$ _____ $= \frac{7}{10}$ b) $\frac{3}{12} +$ _____ $= \frac{9}{12}$ c) _____ $+ \frac{1}{8} = 1\frac{2}{8}$

3 marks

3 Write the missing fractions in these calculations.

a) $\frac{7}{8} -$ _____ $= \frac{2}{8}$ b) _____ $- \frac{2}{7} = \frac{4}{7}$

c) $\frac{8}{10} -$ _____ $= \frac{5}{10}$

3 marks

Marks......... /8

Total marks .... /24   How am I doing?

Fractions (including Decimals)

Finding Fractions

PS Problem-solving questions

Challenge 1

1 Work out:

a) $\frac{1}{3}$ of 15 = _____

b) $\frac{1}{4}$ of 24 = _____

c) $\frac{1}{6}$ of 36 m = _____ m

d) $\frac{1}{8}$ of 32 kg = _____ kg

4 marks

2 Work out:

a) $\frac{2}{3}$ of 24 = _____

b) $\frac{4}{5}$ of 20 = _____

c) $\frac{3}{4}$ of 36 ml = _____ ml

d) $\frac{2}{2}$ of 25 km = _____ km

4 marks

PS 3 Lesa is on a sponsored run. The run is 12 kilometres. Lesa has completed $\frac{1}{6}$ of the course.

How far has she run? _____ km

1 mark

PS 4 There are 28 children in a class. $\frac{1}{2}$ of the class are boys.

How many girls are in the class? _____

1 mark

Marks......... /10

Challenge 2

1 Work out:

a) $\frac{1}{10}$ of 70 = _____

b) $\frac{1}{7}$ of 28 = _____

c) $\frac{1}{12}$ of £24 = £_____

d) $\frac{1}{15}$ of 30 m = _____ m

4 marks

2 Work out:

a) $\frac{2}{5}$ of 35 = _____

b) $\frac{7}{8}$ of 32 = _____

c) $\frac{11}{12}$ of 60 cm = _____ cm

d) $\frac{3}{20}$ of 40 km = _____ km

4 marks

PS 3 Cath has a 40 kilogram bag of coal. She has used $\frac{3}{5}$ of the bag.

How much coal has she used? _____ kg

1 mark

52

Fractions (including Decimals)

Finding Fractions

PS **4** This bar is divided into fifths, but part of the bar is hidden. The value of each fifth is shown.

| 6 | 6 | |

What is the value of the whole bar? _____

1 mark

Marks .... /10

Challenge 3

1 Work out: **a)** $\frac{7}{12}$ of 60 = _____   **b)** $\frac{5}{9}$ of 81 = _____

2 marks

PS **2** A gift shop sells 40 pens. $\frac{2}{5}$ of the pens have blue ink and $\frac{3}{8}$ of the pens have black ink. The rest have red ink.

How many pens have red ink? _____

1 mark

3 Write in the missing numbers.

a) $\frac{3}{4}$ of _____ = 15   **b)** $\frac{3}{10}$ of _____ = 9

2 marks

4 Work out $\frac{3}{2}$ of 10. _____

1 mark

PS **5** There are 27 children in a class. Jo says, "$\frac{1}{2}$ of the class are girls and $\frac{1}{2}$ of the class are boys."

Explain why Jo must be wrong.

_____

_____

1 mark

6 Work out the missing numbers.

a) $\frac{3}{4}$ of _____ = $\frac{1}{8}$ of 24   **b)** $\frac{3}{5}$ of _____ = $\frac{3}{4}$ of 16

2 marks

Marks .... /9

Total marks .... /29   How am I doing?

Fractions (including Decimals)

Fraction and Decimal Equivalents

PS Problem-solving questions

Challenge 1

1 Draw lines to match each fraction to the equivalent decimal.

$\frac{3}{4}$ $\frac{1}{2}$ $\frac{1}{4}$

0.5   0.75   0.25

1 mark

2 Write these fractions as decimals.

a) $\frac{1}{10}$ = _____   b) $\frac{7}{10}$ = _____   c) $\frac{3}{10}$ = _____

3 marks

3 Write as decimals:

a) Nine tenths = _____   b) One and six tenths = _____

c) Two and four tenths = _____   d) Five and five tenths = _____

4 marks

4 Here is a number line.
Write the numbers that the arrows are pointing to.

a) _____   b) _____   c) _____

0 ———————————————————— 1

3 marks

Marks......... /11

Challenge 2

1 Write these fractions as decimals.

a) $\frac{1}{2}$ = _____   b) $\frac{1}{4}$ = _____   c) $\frac{3}{4}$ = _____

3 marks

2 Write these fractions as decimals.

a) $\frac{9}{10}$ = _____   b) $\frac{57}{100}$ = _____   c) $\frac{3}{100}$ = _____

3 marks

Fractions (including Decimals)

Fraction and Decimal Equivalents

3 Gia says, "I have got four pounds and fifty-seven pence."

What is the usual way to write this using digits? _____

1 mark

4 Here is a number line.
Write the decimals that the arrows are pointing to.

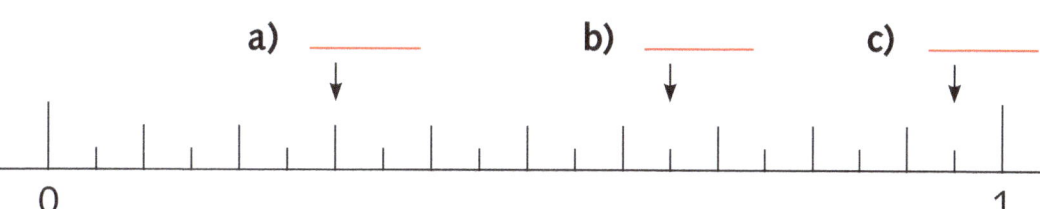

a) _____   b) _____   c) _____

3 marks

Marks......... /10

Challenge 3

1 Write these fractions as decimals.

a) $\frac{9}{100}$ = _____   b) $\frac{20}{100}$ = _____   c) $\frac{101}{100}$ = _____

3 marks

2 Estimate the decimals that the arrows are pointing to.

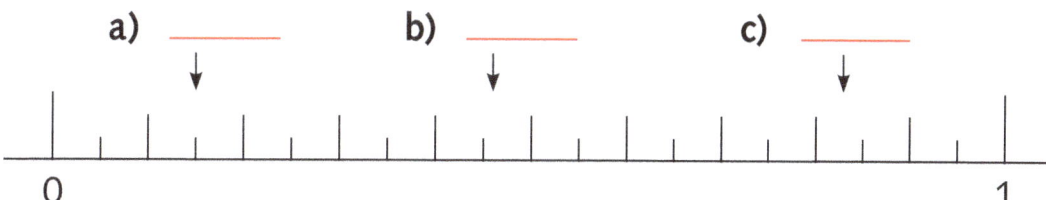

a) _____   b) _____   c) _____

3 marks

3 Look at what James says.
Explain why he is incorrect.

_____
_____
_____

$\frac{7}{100}$ is 0.7

1 mark

4 Write in the missing numbers.

a) $\frac{3}{10}$ = 0.03 × _____   b) $\frac{70}{100}$ = 0.07 × _____

2 marks

Marks.......... /9

Total marks .... /30   How am I doing?

55

Fractions (including Decimals)

Rounding Decimals

PS Problem-solving questions

Challenge 1

1 Round each decimal to the nearest whole number.

a) 5.1 _____   b) 4.8 _____   c) 2.9 _____

d) 6.4 _____   e) 7.6 _____   f) 3.3 _____

6 marks

PS 2 Zain cut a length of wood. It was 2.7 metres long.

Write the length of the wood to the nearest metre. _____ m

1 mark

PS 3 A jug is filled with 1.2 litres of drink.

Write the capacity to the nearest litre. _____ litre(s)

1 mark

PS 4 Kat thinks of a decimal number between 6 and 7. She rounds her number to the nearest whole number. The answer is 7.

Write what Kat's number could have been. _____

1 mark

Marks.......... /9

Challenge 2

1 Round each decimal to the nearest whole number.

a) 7.4 _____   b) 6.5 _____   c) 21.8 _____

d) 53.7 _____   e) 86.3 _____   f) 105.3 _____

6 marks

PS 2 Sam weighs 49.6 kilograms.

Write this mass to the nearest kilogram. _____ kg

1 mark

PS 3 Tiana rounds a number to the nearest whole number. The answer is 50.

Write what Tiana's number could have been.

_____

1 mark

56

Fractions (including Decimals)

Rounding Decimals

PS **4** Debbie rounds numbers to the nearest whole number.
Circle the numbers she will round to 30.

30.6   32.1   29.7   30.4   29.3

2 marks

Marks .... /10

Challenge 3

1 Round each decimal to the nearest whole number.

a) 116.3 _____   b) 229.8 _____   c) 707.0 _____

3 marks

PS **2** Scott rounds numbers to the nearest whole number.
Circle the numbers he will round to 100.

98.9   100.3   99.6   100.9   99.1

2 marks

PS **3** Fay adds 6.2 and 7.4 together.
She estimates the answer by rounding each number to the nearest whole.

Write the total of the rounded numbers. _____

1 mark

PS **4** Aaron rounds numbers to the nearest whole number.
Draw lines to match the decimal numbers to the rounded numbers.

48.7   49.8   50.4   49.5   49.4

49   50

1 mark

Marks .... /7

Total marks .... /26   How am I doing?

Fractions (including Decimals)

Comparing Decimals

PS Problem-solving questions

Challenge 1

1. Circle the largest decimal number.

   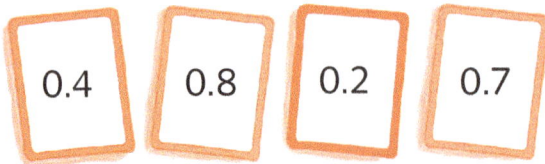
   0.4   0.8   0.2   0.7

   1 mark

2. Circle the smallest decimal number.

   2.4   2.8   3.1   2.5

   1 mark

3. Write these decimal numbers in order, starting with the smallest.

   5.6   **6.2**   **5.9**   **4.7**

   _____   _____   _____   _____

   1 mark

PS 4. George has £3.56, Raja has £5, Craig has £7.32 and Obe has £4.98.

   Write who has the most money.   _____

   1 mark

PS 5. Write a decimal between 7.6 and 8.   _____

   1 mark

   Marks.........../5

Challenge 2

1. Circle the largest decimal in this list.

   5.89   6.25   12.16   8.99   3.75

   1 mark

2. Write these decimals in order, starting with the smallest.

   34.45   **45.43**   **45.34**   **35.44**   **43.54**

   _____ _____ _____ _____ _____

   1 mark

Fractions (including Decimals)

Comparing Decimals

PS 3 Ned puts these decimals in order, starting with the smallest.

60.05   60.56   66.06   65.66   60.06

Write the decimal that is third in Ned's list. _____   *1 mark*

4 Write a decimal between 23.56 and 23.62 _____   *1 mark*

5 Akash orders five numbers. Write what the missing numbers could be.

45.43 > _____ > 45.34 > _____ > 45.27   *2 marks*

Marks.......... /6

Challenge 3

1 Put the correct symbol < or > in each circle.

a) 5.67   6.57   b) 23.56   23.46   *2 marks*

PS 2 Jay has five pieces of wood. They are 3.25 m, 3.76 m, 2.88 m, 3.75 m and 2.98 m long.

a) Write the longest length. _____ m

b) Write the shortest length. _____ m   *2 marks*

3 5.6 and 5.7 are shown on this number line.

Mark 4.8 and 6.2 on the number line.

2 marks

4 Write these numbers in order, smallest first.

52.5   5.25   6   50   25.5

_____ _____ _____ _____ _____

1 mark

Marks.......... /7

Total marks .... /18   How am I doing?

59

Fractions (including Decimals)

Dividing by 10 and 100

PS Problem-solving questions

Challenge 1

1 Complete these calculations.

a) 6 ÷ 10 = _____   b) 12 ÷ 10 = _____

c) 0.7 ÷ 10 = _____   d) 2.5 ÷ 10 = _____

4 marks

2 Fill in the missing numbers.

a) _____ ÷ 10 = 0.5   b) _____ ÷ 10 = 3.6

c) _____ ÷ 10 = 0.63

3 marks

3 Write the value of the 7 in the answer of each calculation.

a) 7 ÷ 10 = _____   b) 70 ÷ 10 = _____

c) 0.7 ÷ 10 = _____

3 marks

PS **4** A 2.5 kg bag of dog food is enough for 10 meals.

How much dog food is used for each meal? _____ kg

1 mark

Marks.......... /11

Challenge 2

1 Complete these calculations.

a) 6.7 ÷ 10 = _____   b) 54 ÷ 100 = _____

c) 15 ÷ 100 = _____   d) 78 ÷ 10 = _____

4 marks

2 Fill in the missing numbers.

a) _____ ÷ 10 = 0.57   b) _____ ÷ 10 = 3.6

c) 92 ÷ _____ = 0.92   d) 46 ÷ _____ = 0.46

4 marks

PS **3** Sunita divides a number by 100. The answer is six tenths.

Write the number Sunita divided. _____

1 mark

Fractions (including Decimals)

Dividing by 10 and 100

4 Write the value of the 4 in the answer of each calculation.

a) 34 ÷ 100 = _____   b) 6.4 ÷ 10 = _____

2 marks

PS 5 Gus buys 100 of the same stamps and pays £64 for them.

What is the cost of each stamp?   £ _____

1 mark

Marks .... /12

Challenge 3

1 Write in the missing numbers.

a) 12 ÷ _____ = 1.2   b) 27 ÷ _____ = 0.27

c) _____ ÷ 10 = 3.2   d) _____ ÷ 100 = 0.79

4 marks

PS 2 Kagendo divides a number and works out the answer to be nine hundredths.
Sally divides the same number and works out the answer to be nine tenths.

Write what Kagendo's and Sally's calculations could have been.

Kagendo's calculation: _____

Sally's calculation: _____

2 marks

3 Circle the calculations that have answers with 6 tenths.

| 64 ÷ 10 | 56 ÷ 100 | 6.7 ÷ 10 | 56 ÷ 10 | 36 ÷ 100 |

2 marks

PS 4 Write division calculations dividing by 10.

a) _____ ÷ _____ = four ones and five tenths

b) _____ ÷ _____ = six ones and eight hundredths

2 marks

Marks .... /10

Total marks .... /33   How am I doing?

Fractions (including Decimals)

Decimal Problems

PS Problem-solving questions

Challenge 1

1 A shop sells the following items.

£4.95 £1.25 £2.75 £5.25

a) Holly buys a set of pens, a ruler and a pencil case. What was the total cost? £_____

b) Josef has a £10 note. He buys a book. How much change does he get? £_____

c) Katie spent £7.70 on two items. Which two items did she buy?

_____   _____

3 marks

2 Michelle has two bottles that each hold 4 litres of water. One is half full and the other is one-quarter full. How any litres of water does Michelle have? _____ litres

1 mark

3 Obi lives 2500 metres from school. He walks to and from school five days a week. How far does he walk in one week walking to and from school? Give your answer in kilometres. _____ km

1 mark

4 Oranges cost £1.20 per kilogram. Jane buys 2 kilograms of oranges. How much does she pay? £_____

1 mark

Marks.......... /6

Challenge 2

1 Pippa has £50. She buys a dress for £19.90 and a pair of shoes for £24.99. How much money does Pippa have left? £_____

1 mark

2 Bananas cost 70p per kilogram. Tim buys $1\frac{1}{2}$ kg of bananas. How much does he pay? £_____

1 mark

Fractions (including Decimals)
Decimal Problems

PS 3 Sammy uses 650 grams of flour from a 2 kilogram bag. How much flour is left in the bag? _____ kg *1 mark*

PS 4 Dominic spends £18.70 on three books. Two of the books cost £5.70 and £4.90. What is the cost of the third book? £_____ *1 mark*

Marks .... /4

Challenge 3

PS 1 A bag of potatoes has a mass of 4.8 kilograms. What is the total mass of $2\frac{1}{2}$ bags? _____ kg *1 mark*

PS 2 A glass holds $\frac{1}{2}$ litre. How many times can the glass be filled from a 5 litre bottle? _____ *1 mark*

PS 3 This is a map showing roads and some towns.

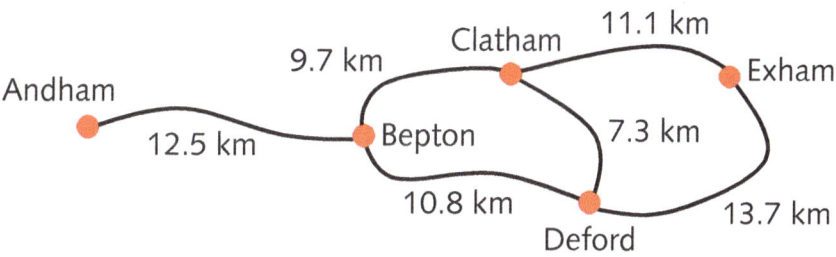

a) Jed drives from Andham to Clatham. How long is his shortest route? _____ km

b) Dimi drives from Bepton to Exham. How long is the shortest route? _____ km

c) Maisie drives from Deford to Exham going through Clatham. She runs out of petrol 3 kilometres from Exham. How far has she driven? _____ km

3 marks

Marks .... /5

Total marks .... /15   How am I doing?

Progress Test 2

PS Problem-solving questions

1. Write the next numbers in the sequences.

   a)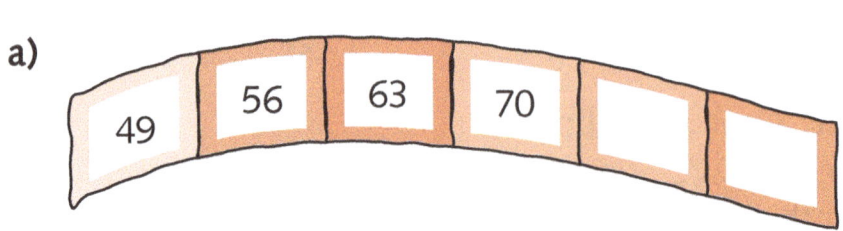

   | 49 | 56 | 63 | 70 | | |

   b)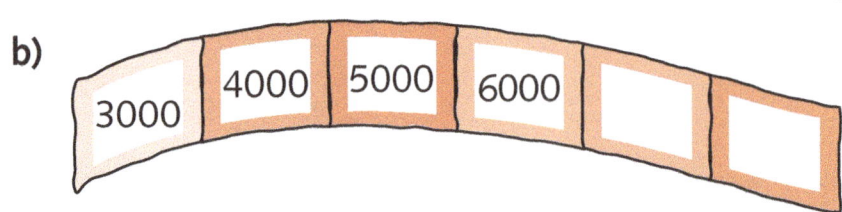

   | 3000 | 4000 | 5000 | 6000 | | |

   2 marks

2. Write these numbers in order, starting with the smallest.

   8314   1834   1384   8341   4138

   ____   ____   ____   ____   ____

   1 mark

3. Circle the value of the digit 1 in the number 6914.

   | 1 one | 1 ten | 1 hundred | 1 thousand |

   1 mark

4. 538 can be written as 400 + 130 + 8.

   Write 538 in two different ways.

   a) _____   b) _____

   2 marks

5. Complete the table by rounding each number to the nearest 10.

Number	Rounded to the nearest 10
3784	
427	
79	

   3 marks

6. Write the next number in the sequence.

   | 18 | 12 | 6 | 0 | |

   1 mark

64

Progress Test 2

7. Work out:

   a) 6734 + 264 = _____   b) 5183 − 437 = _____

   2 marks

8. Estimate the answer to 568 + 357 by rounding the numbers to the nearest hundred and adding the rounded numbers.

   _____ + _____ = _____

   1 mark

PS 9. Josh buys and sells computer games. He has 387 games. He buys 56 games and sells 125 games.

   How many computer games does he have now?   _____ games

   1 mark

PS 10. Chloe has 65 metres of wood. She buys 120 more metres of wood to build a fence. She has 36 metres left over.

   How much wood did Chloe use building the fence? _____ m

   1 mark

11. Work out:

    a) 6 × 4 × 3 = _____   b) 5 × 8 × 6 = _____

    2 marks

12. A factor pair of a number is 56 and 3.

    What is the number? _____

    1 mark

13. Work out:

    a)   5 3 2   b)   9 4 0   c)   6 2 8
       ×   4   ×   8   ×   6
       ───────   ───────   ───────

    3 marks

14. Write the missing numbers in this sentence:

    2 children share 3 paint pots and 5 brushes, so 4 children share

    _____ paint pots and _____ brushes.

    2 marks

15. Write in the missing number: 48 × 3 = _____ × 3 + 8 × 3 = 144

    1 mark

Progress Test 2

PS Problem-solving questions

16. Work out:

a) 3 ÷ 100 = _____

b) 9 ÷ 100 = _____

c) $\frac{7}{10}$ ÷ 10 = _____

3 marks

17. Estimate the fraction each arrow points to.

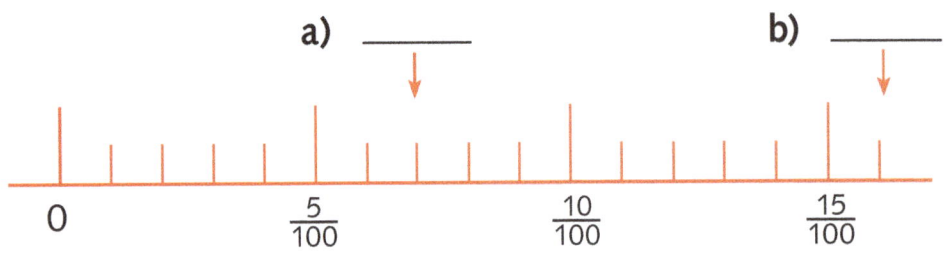

2 marks

18. a) $\frac{6}{8}$ of this circle is shaded. Write in the missing number.

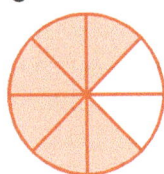

$\frac{\Box}{4}$ of this circle is shaded.

b) $\frac{1}{2}$ of this rectangle is shaded. Write in the missing number.

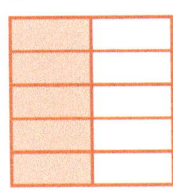

$\frac{\Box}{10}$ of this rectangle is shaded.

2 marks

19. Work out:

a) $\frac{4}{10} + \frac{7}{10}$ = _____

b) $\frac{7}{8} + \frac{2}{8}$ = _____

2 marks

20. Write these decimals as fractions.

a) 0.13 = _____   b) 0.9 = _____   c) 0.09 = _____

3 marks

21. Write these fractions as decimals.

a) $\frac{93}{100}$ = _____   b) $\frac{7}{100}$ = _____   c) $\frac{4}{10}$ = _____

3 marks

Progress Test 2

22. Round each decimal to the nearest whole number.

  a) 4.7 _____ b) 15.4 _____ c) 29.8 _____

3 marks

23. Billy rounds a number, less than 25, with one decimal place to the nearest whole number. The rounded number is 25.

What could Billy's number have been? _____

1 mark

24. Circle the largest number.

73.8   9.3   20.0   68.9   73.7

1 mark

25. Write these numbers in order, starting with the smallest.

31.6   31.8   32.5   30.7   30.2

_____   _____   _____   _____   _____

1 mark

26. Write in the missing numbers.

  a) 5.6 ÷ _____ = 0.56 b) 9 ÷ _____ = 0.09

  c) 16 ÷ _____ = 0.16 d) 0.8 ÷ _____ = 0.08

4 marks

27. Chandra divides a two-digit number by 10. The answer has 5 ones and 7 tenths.

What was the number that Chandra divided? _____

1 mark

28. Work out:

  a) $\frac{2}{5}$ of 25 = _____ b) $\frac{3}{4}$ of 32 = _____

  c) $\frac{9}{10}$ of 90 = _____

3 marks

29. Jade buys 4 t-shirts that cost £7.25 each.

Work out the total cost of the 4 t-shirts. £ _____

1 mark

Marks........./54

Measurement

Comparing Measures

PS Problem-solving questions

Challenge 1

1 Write these lengths in order, longest first.

36.87 m   63.78 m   83.76 m   83.67 m   38.76 m

_____   _____   _____   _____   _____

2 Write these capacities in order, smallest first.

45.78 litres   4.87 litres   5.47 litres   54.87 litres   78.54 litres

_____   _____   _____   _____   _____

PS 3 Sonya has three different-sized jugs. The largest jug holds 4.75 litres and the smallest jug holds 4.56 litres.

What could the capacity of the third jug be? _____ litres

4 There are three lengths of ribbon. The lengths are $1\frac{1}{2}$ metres, 1.25 metres and 1.65 metres.

Write the lengths in order, smallest first.

_____ m   _____ m   _____ m

PS 5 Stan weighs 43.76 kg, Dubry weighs 46.73 kg and Tom weighs 45.87 kg.

Who is the heaviest? _____

Marks.......... /5

Challenge 2

1 Circle the larger mass.

a) 3.2 kg   or   2800 g

b) 7.8 kg   or   8000 g

2 Circle the shorter length.

a) 5.6 m   or   474 cm

b) 67 mm   or   6 cm

3 Circle the larger capacity.

a) 5.675 litres   or   570 ml

b) 8000 ml   or   10 litres

68

Measurement

Comparing Measures

4 Write these lengths in order, shortest first.

475 cm   5 m   4800 mm   4.7 m

_____ _____ _____ _____

1 mark

PS 5 Alvin has a 5 kilogram bag of flour. He uses 255 grams. Has he used over half of the bag? _____

1 mark

Marks .... /8

Challenge 3

1 Circle the longest length.

$4\frac{1}{2}$ m   4.6 m $4\frac{3}{4}$ m   4.8 m   4.2 m

1 mark

2 Circle the shortest length.

5.7 kilometres   4500 metres

4.85 kilometres   5000 metres

1 mark

3 Amy is offered four drinks. She wants the largest drink. Circle the drink she should choose.

300 ml can   0.5 litre bottle   0.25 litre glass   1 litre bottle

1 mark

PS 4 There are three parcels. Parcel A weighs 6 kg, parcel B is $\frac{1}{2}$ kg heavier than parcel A and parcel C is 250 g lighter than parcel B.

Write the parcels in order, the heaviest mass first.

_____ _____ _____

1 mark

Marks .... /4

Total marks .... /17   How am I doing?

Measurement

Estimating Measures

PS Problem-solving questions

Challenge 1

1 Estimate the mass of the parcel.

_____ kg

1 mark

2 Estimate the length of the line below. _____ cm

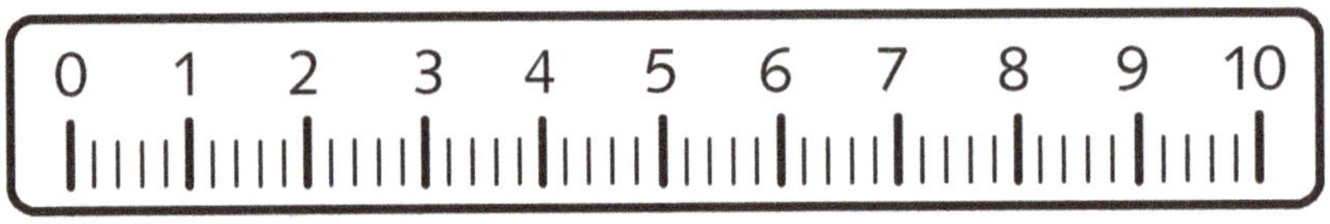

_____

1 mark

3 Estimate how much liquid is in the jug. _____ litre(s)

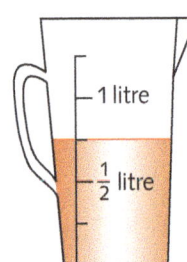

1 mark

4 Phil gets £50 to go shopping. At the end he says, "I have about half of my money left."

Tick the amounts of money Phil could have.

£45   £24.75   £36.50   £26.15   £0.50

2 marks

Marks.......... /5

Challenge 2

1 Tick the best estimate for the length of a car.

| 40 mm | 40 cm | 4 m | 0.4 km |

1 mark

2 Tick the best estimate for the capacity of a mug.

35 ml   35 litres   350 ml   3.5 litres

1 mark

3 Tick the best estimate for the mass of a banana.

| 120 g | 12 g | 1.2 kg | 12 kg |

1 mark

Measurement

Estimating Measures

4 Melanie walks 1 kilometre to school with no stops. Tick the best estimate for the time it will take.

- 1 minute
- 12 minutes
- 45 minutes
- 1 hour

1 mark

Marks........../4

Challenge 3

1 Adam goes for a meal. He chooses the following items.

£7.60   £3.65   £2.55

Estimate the cost of the meal by rounding the prices to the nearest pound and adding the rounded prices.   £ _____

1 mark

2 Estimate the temperature shown on the thermometer.

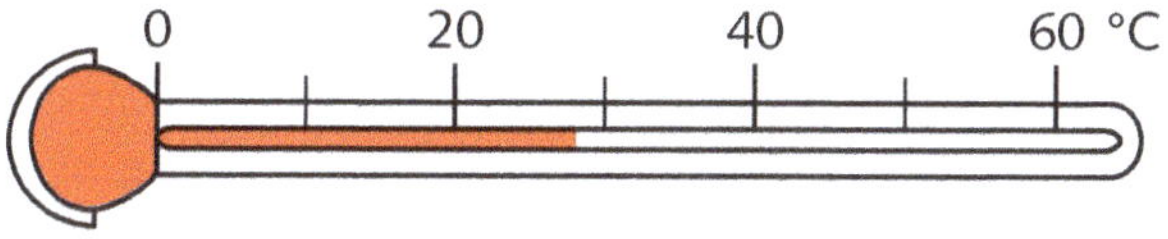

_____ °C

1 mark

3 Estimate the length of the curved line. Do not measure the line.

_____ cm

1 mark

PS 4 The same parcel is shown on both sets of scales. Estimate the mass of the parcel.   _____ kg

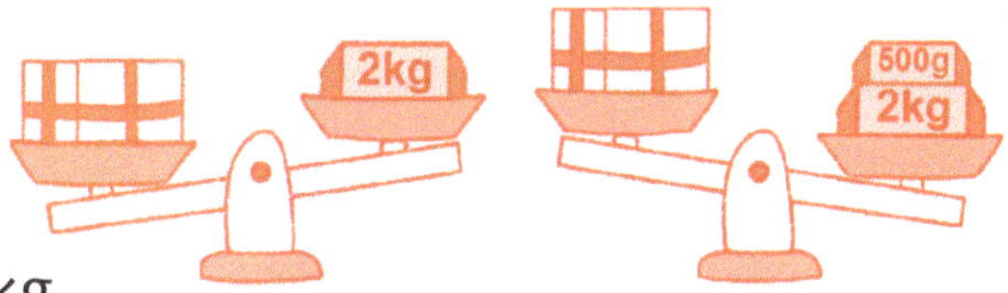

1 mark

Marks........../4

Total marks ..../13   How am I doing?

Measurement
Converting Measures

PS Problem-solving questions

Challenge 1

1 Convert these measurements.

Length
a) 30 mm = _____ cm
b) 400 cm = _____ m
c) 5000 m = _____ km
d) 6 cm = _____ mm
e) 5 m = _____ cm
f) 9 km = _____ m

Mass
g) 4000 g = _____ kg
h) 40 000 g = _____ kg
i) 5 kg = _____ g

Capacity
j) 8 litres = _____ ml
k) 10 000 ml = _____ litres
l) 0.5 litres = _____ ml

12 marks

PS **2** A glass holds 500 ml. How many times can the glass be filled from a 2 litre bottle of drink? _____

1 mark

PS **3** Tina is fitting a kitchen cupboard. The cupboard is 600 mm wide and the space for it is 65 cm.

a) Will the cupboard fit in the space? _____

b) What is the difference between the two measurements? _____

2 marks

Marks......... /15

Challenge 2

1 Convert these measurements.

a) 5.2 m = _____ cm
b) $3\frac{1}{2}$ kg = _____ g
c) 3500 ml = _____ litres
d) 78 mm = _____ cm
e) 250 m = _____ km
f) $\frac{3}{4}$ m = _____ cm
g) 845 cm = _____ m
h) 10 500 m = _____ km

8 marks

Measurement

Converting Measures

PS 2 Sara buys a bag of flour that weighs 1 kilogram. She uses some flour and there are 600 grams left. How much flour has Sara used? _____

1 mark

PS 3 Casey has three jugs. Jug A holds 200 ml, jug B holds 250 ml and jug C holds 500 ml. Which jug will he fill exactly four times to fill a bowl with 1 litre of water? _____

1 mark

PS 4 Ewan has a carton with 568 millilitres of milk. He uses 0.5 litres in a recipe. How much milk does he have left over? _____

1 mark

Marks......... /11

Challenge 3

1 How many millimetres are there in 1 metre? _____ mm

1 mark

2 Write in the missing units.

a) 4.5 km = 4500 _____   b) 67 mm = 6.7 _____

c) 125 m = 0.125 _____   d) 6750 ml = 6.75 _____

e) 5.4 kg = 5400 _____   f) 4.5 m = 450 _____

6 marks

PS 3 Each row and each column adds up to 1 litre. Fill in the missing capacities.

½ litre	300 ml	
		0.6 litres
¼ litre		

1 mark

Marks......... /8

Total marks .... /34   How am I doing?

Measurement

Measurement Calculations

PS Problem-solving questions

Challenge 1

PS 1 Charlotte spends £3.45 on a new pen. How much change will she get from £5?   £ _____

PS 2 Nadia buys a book for £8.49 and a magazine for £4.75. How much does she spend?   £ _____

PS 3 Tomas buys three pens that cost £2 each. How much change will he get from £10?   £ _____

PS 4 Subtract 5p from £5.   £ _____

PS 5 Milly has a £10 note and a £5 note. She spends £12.33. How much money will Milly have left?   £ _____

Marks .... /5

Challenge 2

1 Calculate these measures.

a)   £ 2 . 4 5
   + £ 4 . 6 5
   _____

b)   1 4 . 8 5 m
   − 1 1 . 0 9 m
   _____

c)   5 6 . 7 8 kg
   − 3 8 . 5 8 kg
   _____

d)   3 8 . 9 5 litres
   +   7 . 8 4 litres
   _____

PS 2 Claire goes out for a meal. She spends £15.75 on her main course and £5.50 on dessert. She has £15.35 left. How much did Claire have when she went out?   £ _____

PS 3 Rachael buys two bags of dog food. Each bag weighs 5 kg. Seven days later, she has used 3.5 kg. How much dog food does Rachael have left?   _____ kg

Measurement Calculations

PS 4 At a Sports Day, Mahmood makes three jumps of 3.15 m, 2.87 m and 3.08 m. He adds the distances together. He is 0.15 m short of the school record. What is the school record? _____ m

1 mark

PS 5 David buys some bottles of juice for a party. He buys five 2 litre bottles and ten 0.5 litre bottles. How much juice does he have altogether? _____ litres

1 mark

Marks.......... /8

Challenge 3

PS 1 Asel uses some weights to find the mass of a parcel. She uses two 1 kilogram weights, three 500 gram weights and one 50 gram weight. What is the mass of the parcel?

a) Give your answer in kilograms. _____ kg

b) Give your answer in grams. _____ g

2 marks

PS 2 Toby has £7.25 in his money box. He counts the coins; he has these coins. One coin is missing.

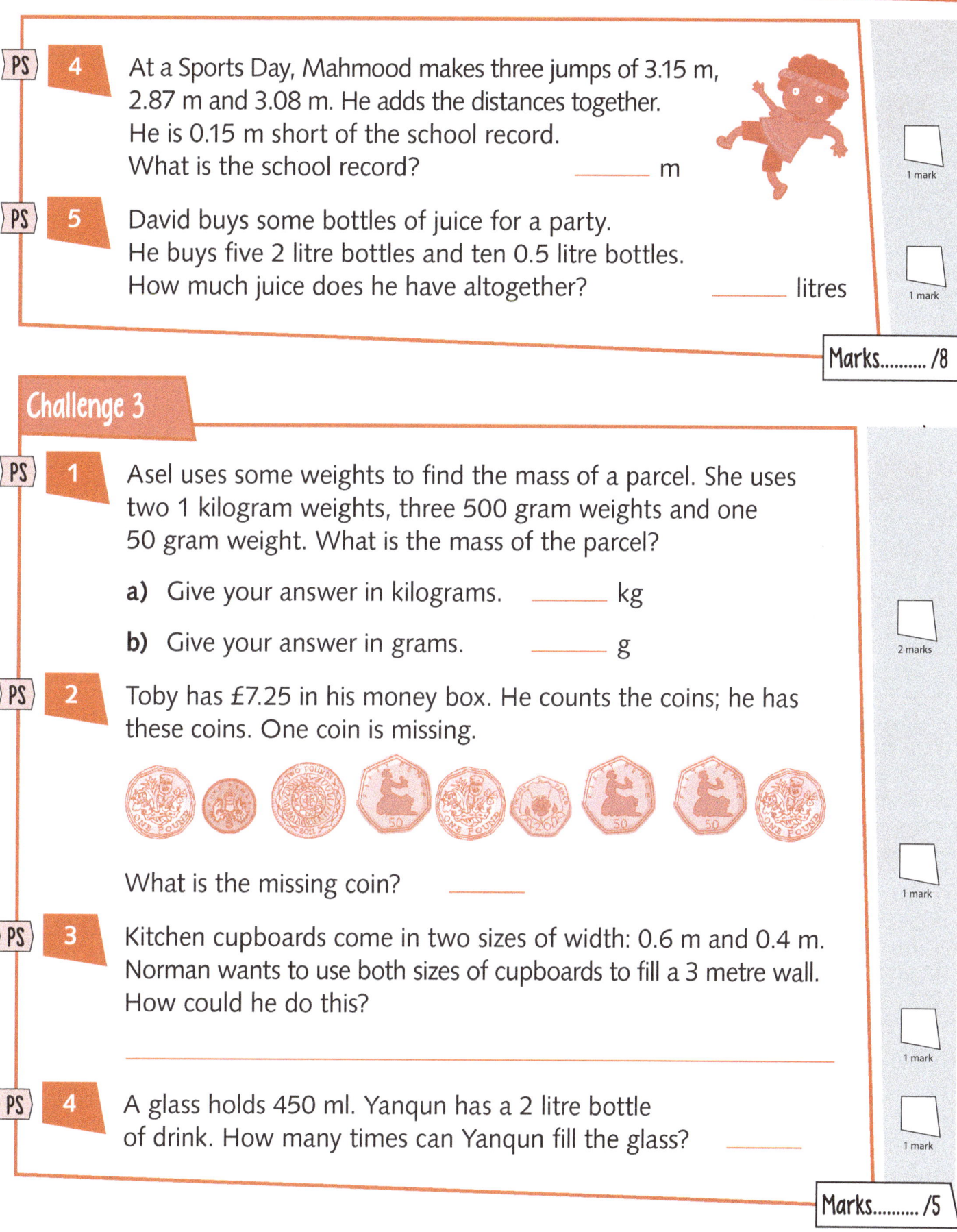

What is the missing coin? _____

1 mark

PS 3 Kitchen cupboards come in two sizes of width: 0.6 m and 0.4 m. Norman wants to use both sizes of cupboards to fill a 3 metre wall. How could he do this?

_____

1 mark

PS 4 A glass holds 450 ml. Yanqun has a 2 litre bottle of drink. How many times can Yanqun fill the glass? _____

1 mark

Marks.......... /5

Total marks .... /18   How am I doing?

Measurement

12 and 24 Hour Time

PS Problem-solving questions

Challenge 1

1 These clocks show morning times.
Write the times shown on the clocks in 12 hour time.

a) _____ b) _____ c) _____

3 marks

2 These clocks show afternoon and evening times.
Write the times shown on the clocks in 24 hour time.

a) _____ b) _____ c) _____

3 marks

3 Write the time shown on these clocks in words.

a) _____ b) _____ c) _____

3 marks

Marks.........../9

Measurement

12 and 24 Hour Time

Challenge 2

1. Write these digital times in 12 hour time.

   a) 15:25 _____   b) 18:10 _____

   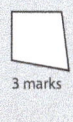 2 marks

2. Write these word times in 24 hour time.

   a) Quarter past two in the afternoon _____

   b) Ten to twelve in the morning _____

   c) Twenty-five to seven in the evening _____

    3 marks

PS 3. The time on this clock is 40 minutes slow.

   What is the correct time? _____

   1 mark

Marks.......... /6

Challenge 3

1. Write the missing parts of each time in this table.

	Word time		12 hour time		24 hour time
a)	Five past two in the afternoon	=	____:05 ____	=	____:05
b)	Twenty-five to ____ in the evening	=	8:____ ____	=	____:____

    2 marks

PS 2. Harpreet's train leaves at 15:25.
She looks at her watch.
How many minutes is it before her
train leaves? _____ minutes

    1 mark

Marks.......... /3

Total marks .... /18   How am I doing?

Measurement

Time Problems

PS Problem-solving questions

Challenge 1

1 Write in the missing numbers.

a) 2 minutes = _____ seconds   b) 2 hours = _____ minutes

c) 2 days = _____ hours   d) 2 weeks = _____ days

4 marks

2 Write the four months of the year that have exactly 30 days.

_____   _____

_____   _____

4 marks

3 If the year 2020 is a leap year, which year will be the next leap year? _____

1 mark

PS 4 Kim's birthday is on Tuesday, 4th May.
She has her birthday party on the first Saturday after her birthday.

What is the date of the birthday party? _____

1 mark

Marks......... /10

Challenge 2

PS 1 A film starts at 16:45 and finishes at 19:05.

How long does it last? _____ hours _____ minutes

1 mark

PS 2 Carly gets up when her alarm clock shows 07:30.
She gets washed and dressed and has her breakfast. Her watch shows the time that she leaves for school.

How long has it taken her to get ready? _____ minutes

1 mark

PS 3 It is 18:30 and Bev says, "I'll see you in three-quarters of an hour."

What will the time be? _____

1 mark

78

Measurement

Time Problems

4 Change these units of time.

a) 4 weeks = _____ days   b) 2 years = _____ months

c) 5 minutes = _____ seconds   d) 3 days = _____ hours

4 marks

Marks………/7

Challenge 3

PS 1 These time periods are in order, shortest first.
Write the number of minutes that could be the missing number.

| 480 seconds | _____ minutes | $\frac{1}{6}$ of an hour |

1 mark

2 Write in the missing numbers.

a) _____ hours = _____ days = 2 weeks

b) _____ months = _____ years = 1 decade

2 marks

PS 3 Barry goes to sleep at 21:30 and he sleeps for $9\frac{1}{4}$ hours.

At what time will he wake up? _____

1 mark

PS 4 School starts at 8:50am and ends at 12:15pm. The children have a 20-minute break. The rest of the time they are in the classroom.

For how long are they in the classroom?

_____ hours _____ minutes

1 mark

PS 5 Karen catches a train at 10:15am.
The journey lasts 3 hours 12 minutes.

At what time does the train arrive? _____

1 mark

Marks………/6

Total marks …………/23   How am I doing?

Measurement

Perimeter and Area

Challenge 1

1 Three rectangles and squares (**A–C**) are drawn on a centimetre square grid. Find **a)** the area and **b)** the perimeter of each shape.

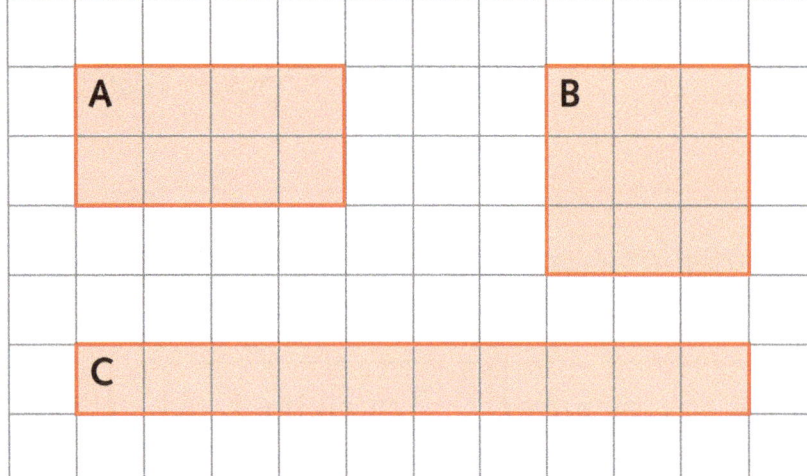

A a) _____ squares
   b) _____ cm

B a) _____ squares
   b) _____ cm

C a) _____ squares
   b) _____ cm

Grid not to scale

6 marks

Marks .... /6

Challenge 2

1 Three shapes (**A–C**) are drawn on a centimetre square grid. Find **a)** the area and **b)** the perimeter of each shape.

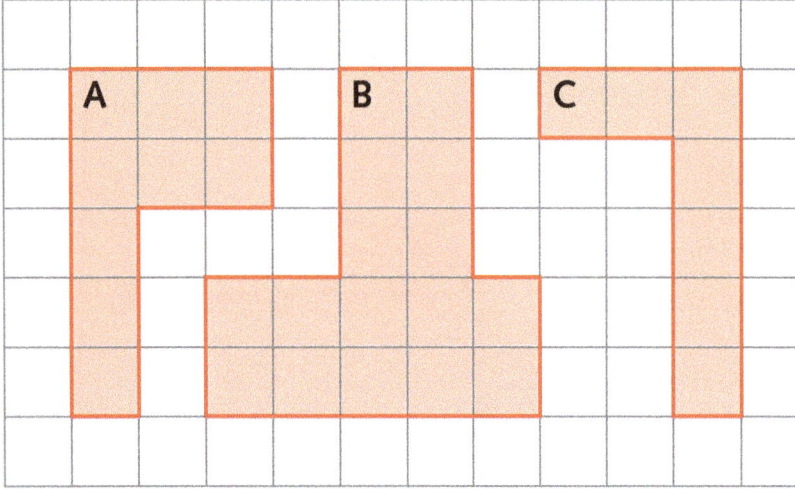

A a) _____ squares
   b) _____ cm

B a) _____ squares
   b) _____ cm

C a) _____ squares
   b) _____ cm

Grid not to scale

6 marks

Measurement

Perimeter and Area

2 Work out the perimeter of this shape. Use a ruler. _____ cm

1 mark

Marks........../7

Challenge 3

1 **a)** The perimeter of this rectangle is 14 cm.

What is the width of the rectangle?

_____ cm

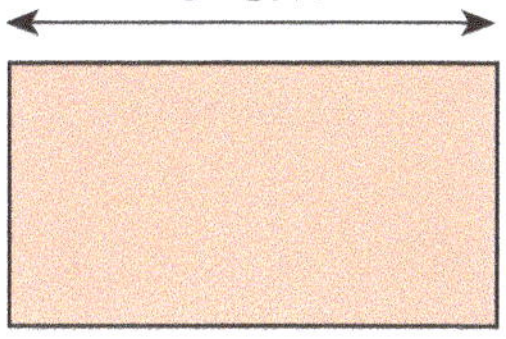

5 cm

Not to scale

b) 25 small squares have been used to make this larger square.

How many small squares fit along one side of the large square?

_____ small squares

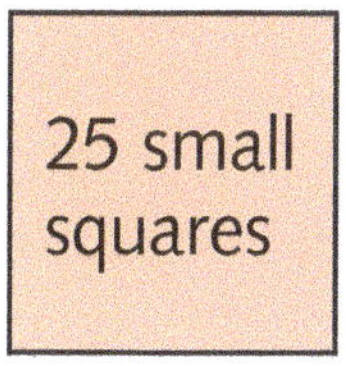

25 small squares

2 marks

2 Bill uses rectangles like this.
They are 2 cm long and 1 cm wide.

He made the design on the right using three rectangles.
What is the perimeter of Bill's design?

_____ cm

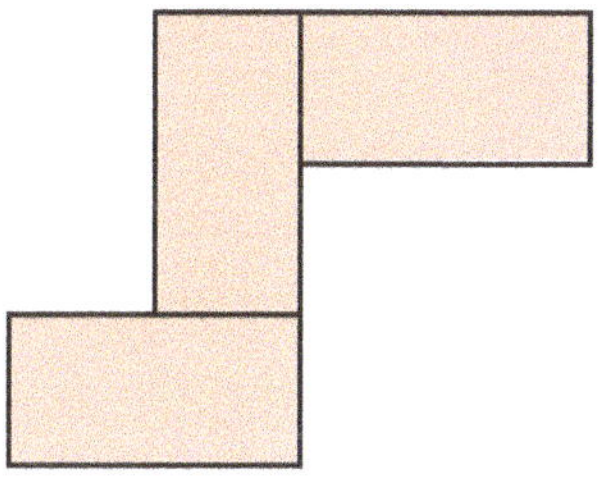

1 mark

Marks........../3

Total marks ..../16   How am I doing?

Progress Test 3

PS Problem-solving questions

1. Write in the missing numbers.

   a) 63 _____ 81 _____ 99 108

   b) 36 _____ _____ 54 60 66

   2 marks

2. Write these numbers in order, largest first.

   3572   3752   3725   3727   3527

   _____ _____ _____ _____ _____

   1 mark

PS 3. Write a four-digit number with 8 ones, 5 hundreds, 6 thousands and 4 tens. _____

   1 mark

4. Circle the Roman numeral that is the largest number.

   XXVIII   LXV   XL   XXXIX

   1 mark

5. Complete the table by rounding these numbers to the nearest 100.

Number	Rounded to the nearest hundred
7 8 2 3	
7 8 2	
7 8	

   3 marks

PS 6. Each day a school kitchen prepares 240 hot meals, 180 salads and 125 vegetarian meals.

   a) How many meals are prepared each day? _____

   b) On Monday, 43 meals are left uneaten. How many meals are eaten on Monday? _____

   2 marks

7. Write in the missing numbers.

   a) 6 × _____ = 36   b) 48 ÷ _____ = 8

   c) _____ × 9 = 45   d) _____ ÷ 7 = 7

   4 marks

82

8. Fill in the missing number.   6 × 5 × _____ = 240

9. Work out:

   a)   5 0 9   b)   4 5 2   c)   8 6 1
      ×   9   ×   7   ×   6
      _____   _____   _____

      _____   _____   _____

10. It takes Alice 2 hours to build 7 metres of fence.
    How long will it take her to build 21 metres of fence? _____ hours

11. Write in the missing numbers.

    a) 6 ÷ _____ = $\frac{6}{100}$ b) 8.1 ÷ _____ = $\frac{81}{100}$

    c) 0.7 ÷ _____ = $\frac{7}{100}$

12. Draw a line from $\frac{3}{5}$ to its equivalent fraction.

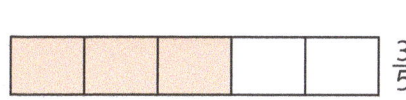

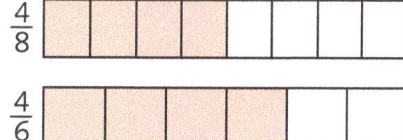

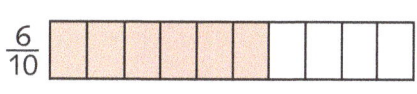

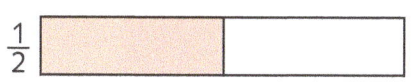

13. Fill in the missing numbers.

    a) $\frac{\square}{8} + \frac{6}{8} = \frac{7}{8}$ b) $\frac{\square}{12} + \frac{5}{12} + \frac{3}{12} = \frac{11}{12}$

14. Heather has saved up £18.70.
    Write this to the nearest whole pound.   £ _____

15. Mo has £20.
    She spends $\frac{1}{5}$ of her money on a drink and $\frac{3}{10}$ on a sandwich.
    How much money does she have left?   £ _____

Progress Test 3

PS Problem-solving questions

PS **16.** Kate buys some things for her party.

- Balloons £1.75
- Candles £3.85
- Cake £11.85
- Plates £5.80
- Mugs £4.80

a) She spends exactly £17.45. Which three things did she buy?

_____   _____   _____

b) Kate pays with a £20 note.
How much money does she have left?   £ _____

2 marks

17. Write these amounts in order, starting with the largest.

£3   90p   £0.08   £1.25   5p

_____   _____   _____   _____   _____

1 mark

PS **18.** Harvey spends £8, Tia spends 675p, Christopher spends £9.89 and Raf spends two £5 notes.

a) Who spends the least?   _____

b) Who spends the most?   _____

2 marks

PS **19.** Daryl estimates how much money he has spent. He paid £17.45 for a ticket to a theme park, £8.40 for food and £15.35 on gifts.

He rounds each amount to the nearest whole pound and adds the rounded amounts.

Write Daryl's estimate.   £ _____

1 mark

PS **20.** Katherine says, "My train leaves at quarter to 9."
The clock in the station shows 08:30.

How many minutes does Katherine need to wait?   _____ minutes

1 mark

21. The clock on the right shows an evening time.
Circle the digital clock that shows the same time.

A 20:40   B 19:40   C 21:35   D 20:50

1 mark

Progress Test 3

22. Tick the correct statement.

  A There are 300 minutes in 5 hours. ☐

  B There are 60 minutes in 5 hours. ☐

  C There are 35 minutes in 5 hours. ☐

  D There are 120 minutes in 5 hours. ☐

23. Susie runs a lap of a track in 100 seconds.

  Write this as minutes and seconds. _____ minutes _____ seconds

24. Change these measurements.

  a) 5 kg = _____ g   b) 60 mm = _____ cm

  c) 1000 cm = _____ m   d) 8000 ml = _____ litres

  e) 8 m = _____ cm   f) 2 km = _____ m

25. Measure the perimeter of each shape.

  a) _____ cm   b) _____ cm

26. Write the area of the shaded shape. _____ squares

Marks......../46

Geometry – Properties of Shapes

2-D Shapes

Challenge 1

1 Tick the shapes that are rectangles.

A  B  C  D

2 marks

2 These triangles are drawn on a squared grid.

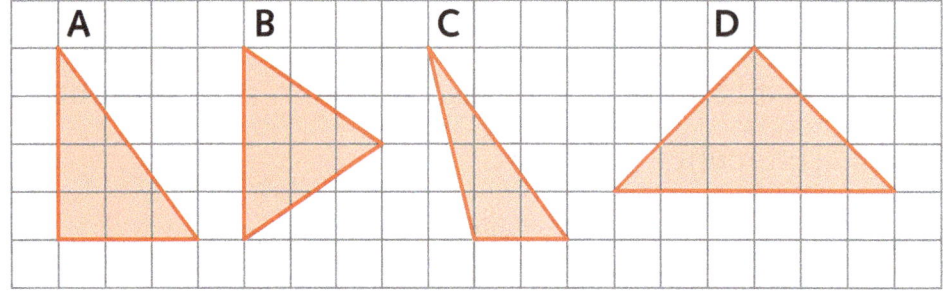

a) Write the letters of the two isosceles triangles. _____

b) Write the letters of the two right-angled triangles. _____

c) Write the letter of the obtuse-angled triangle. _____

d) Write the letter of the acute-angled triangle. _____

4 marks

3 Tick the shape that is a parallelogram.

A  B  C  D

1 mark

Marks.......... /7

Challenge 2

1 Tick the quadrilaterals that are squares.

A  B  C  D

2 marks

Geometry – Properties of Shapes

2-D Shapes

2 Give a reason why this shape is not a hexagon.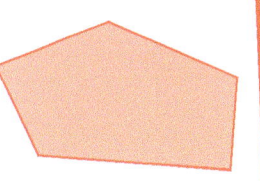

_____

_____

1 mark

3 These triangles are drawn on a squared grid.
Tick the triangle that is a right-angled isosceles triangle.

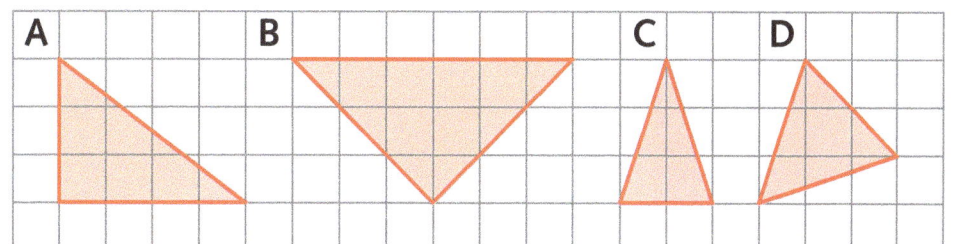

1 mark

Marks......... /4

Challenge 3

1 Circle the shape that has the most sides.

         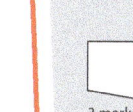

1 mark

2 Tick the shapes with pairs of equal and parallel sides.

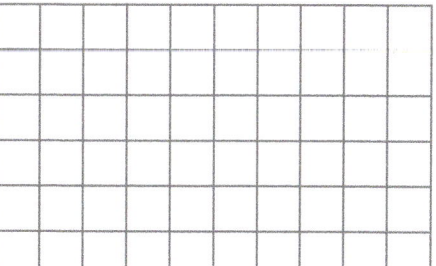

3 marks

3 Mark says, "A pentagon cannot have a right angle."
Draw a pentagon to show that Mark is incorrect.

1 mark

Marks......... /5

Total marks .... /16   How am I doing?

87

Geometry – Properties of Shapes

3-D Shapes

PS Problem-solving questions

Challenge 1

1. Tick the cube.

   A    B 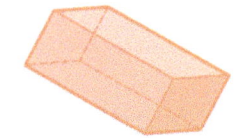   C    D

   1 mark

2. Tick the cone.

   A    B    C    D

   1 mark

3. How many faces does each of these shapes have?

   a)  _____   b) A cube _____

   2 marks

Marks........../4

Challenge 2

1. How many faces does each of these shapes have?

   a) 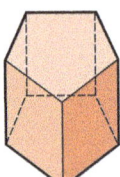 _____   b) 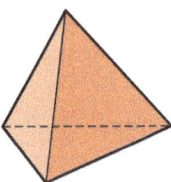 _____

   2 marks

2. Tick the correct statement.

   A Cuboids do not have faces that are perpendicular and parallel.

   B Cuboids only have faces that are perpendicular, and not parallel.

   C Cuboids only have faces that are parallel, and not perpendicular.

   D Cuboids have faces that are perpendicular and parallel.

   1 mark

Geometry – Properties of Shapes

3-D Shapes

3 Complete this table.

Shape	Faces	Edges	Vertices
Cuboid			
Square-based pyramid			
Triangular prism			

3 marks

Marks......... /6

Challenge 3

1 Shona says, "The faces on a cuboid are always rectangles."
Is Shona correct? Explain your answer.

_____

_____

_____

1 mark

2 A prism has 15 edges. What shape is its end face?

_____

1 mark

3 a) How many cubes have been used to make this shape? _____

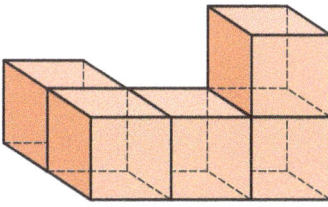

b) Rohan adds more cubes to this shape to make a cuboid. The cuboid is 3 cubes long, 2 cubes wide and 2 cubes high.

How many more cubes will Rohan need? _____

2 marks

Marks......... /4

Total marks .... /14   How am I doing?

Geometry – Properties of Shapes

Lines of Symmetry

Challenge 1

1 Draw the lines of symmetry on these shapes.

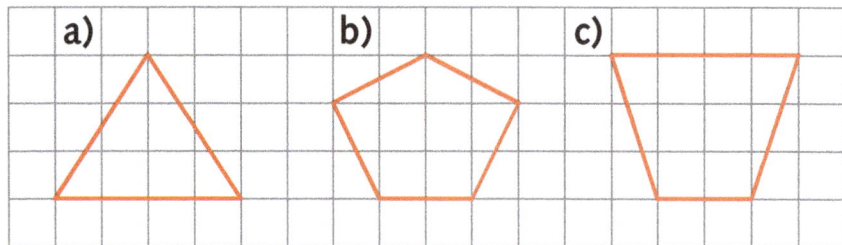

3 marks

2 The dotted lines are lines of symmetry. Complete the shapes.

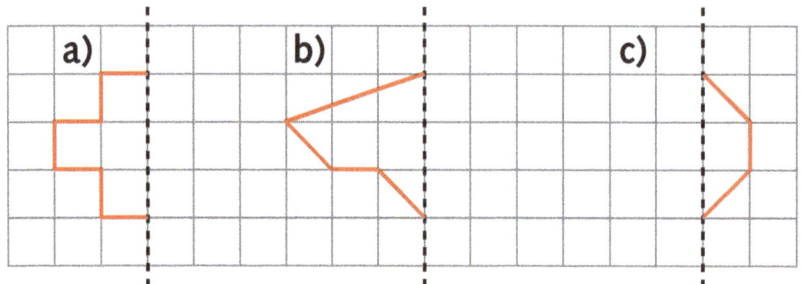

3 marks

Marks......... /6

Challenge 2

1 How many lines of symmetry does each of these shapes have?

a)  _____

b)  _____

c)  _____

3 marks

Geometry – Properties of Shapes

Lines of Symmetry

2 The dotted lines are lines of symmetry. Complete the shapes.

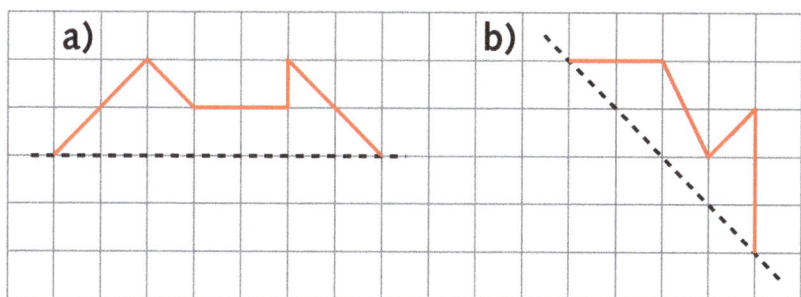

2 marks

Marks......... /5

Challenge 3

1 Tick the shape that has no lines of symmetry.

A B C D E

 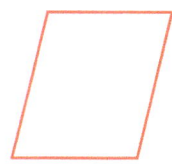 

1 mark

2 Draw three more lines on the grid to make a pentagon with one line of symmetry.

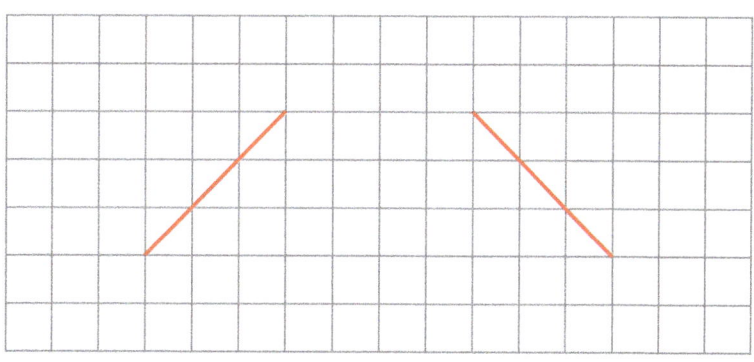 

1 mark

Marks......... /2

Total marks .... /13   How am I doing?

Geometry – Properties of Shapes

Angles

PS Problem-solving questions

Challenge 1

1 Write the number of angles in each shape.

a) Rectangle　　b) Hexagon　　c) Kite

3 marks

2 Tick the acute angle.

A　　B　　C　　D

1 mark

3 Tick the obtuse angle.

A　　B　　C　　D

1 mark

4 Tick the largest angle.

A　　B　　C　　D

1 mark

Marks.......... /6

Challenge 2

1 Here are two shapes drawn on a squared grid.
Tick the acute angles in each shape.

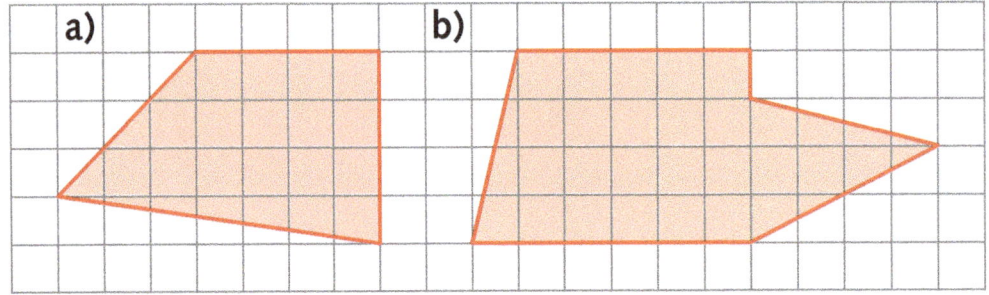

2 marks

Geometry – Properties of Shapes

Angles

2 Write the letters of these angles in order of size, smallest first.

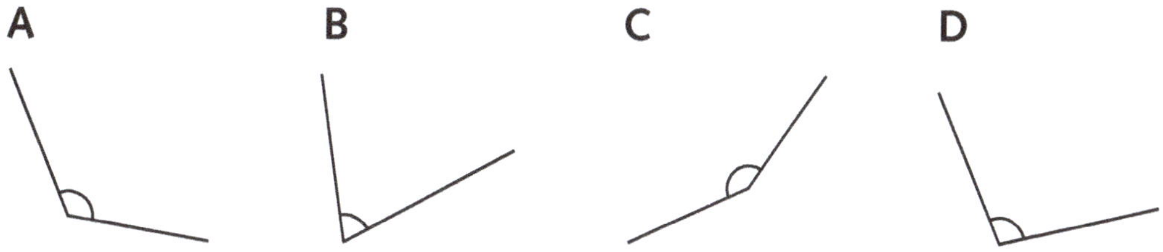

A   B   C   D

_____   _____   _____   _____

3 Write these angles in order of size, starting with the largest.

right   obtuse   acute

_____   _____   _____

Marks.......... /4

Challenge 3

1 Below is a squared grid.

   a) Draw a five-sided shape with three obtuse angles.

   b) Draw a four-sided shape with three acute angles.

PS 2 Caroline says, "Any 2-D shape can have two right angles."
Is Caroline correct? Give a reason for your answer.

_____

_____

Marks.......... /3

Total marks .... /13   How am I doing?

Geometry – Position and Direction

Translations

Challenge 1

1 A square is drawn in different positions on a grid.

Listed below are translations of the squares. Write the letter of the position the square moves to.

a) Square A moves 3 right and 2 down. _____

b) Square F moves 7 right and 2 up. _____

c) Square B moves 5 left and 4 down. _____

d) Square E moves 5 left and 3 up. _____

e) Square C moves 3 right and 2 up. _____

f) Write the letters of the squares that show the translation 2 right and 2 down. Square _____ to square _____.

6 marks

Marks.......... /6

Challenge 2

1 A rectangle is drawn in different positions on a grid.

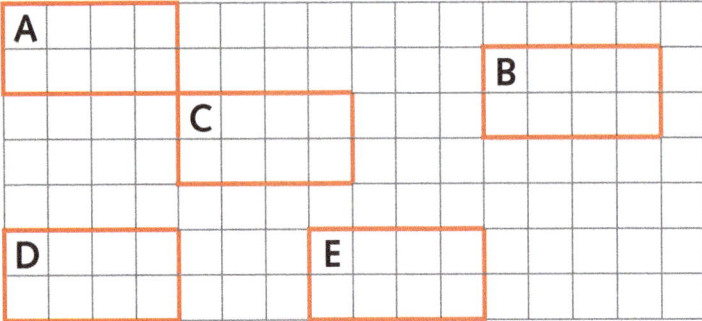

Describe these translations of the rectangle.

a) From A to B _____

94

Geometry – Position and Direction

Translations

b) From E to C _____

c) From B to C _____

d) From D to B _____

e) From A to C _____

f) From C to D _____

6 marks

Marks .... /6

Challenge 3

1 A triangle is translated 5 units left and 7 units down. Write the translation that moves the triangle to its original position.

_____

1 mark

2 Here is a square drawn on a grid.

a) Translate the square 4 right and 2 up. Draw the square.

b) Translate the square 1 left and 1 up. Draw the square.

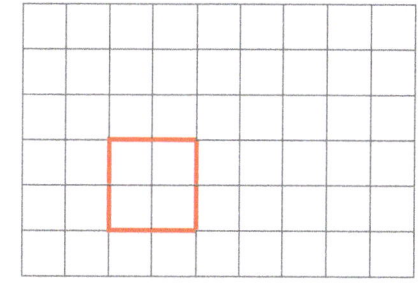

2 marks

3 Circle the correct bold word in each statement.

a) A triangle is translated 6 units left and 0 units down.

This will leave the triangle in the same **row** / **column**.

b) A rectangle is translated 0 units right and 5 units up.

This will leave the rectangle in the same **row** / **column**.

2 marks

Marks .... /5

Total marks .... /17   How am I doing?

Geometry – Position and Direction

Coordinates

Challenge 1

1 Look at the coordinate grid. Write the coordinates of these symbols.

a) ★ _____   b) ♥ _____

c) ⚡ _____   d) ✚ _____

e) ▲ _____   f) ☾ _____

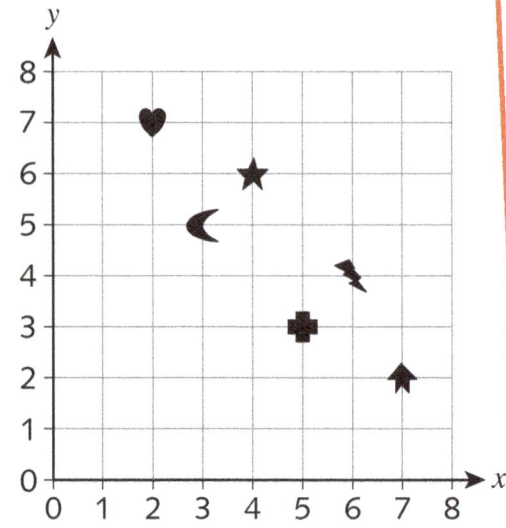

6 marks

2 Look again at the coordinate grid.

a) Plot these points on the grid: (1, 2) and (7, 8).
Join the points with a straight line.

b) What are the coordinates of the mid-point of the line? _____

2 marks

Marks.......... /8

Challenge 2

1 Plot these points on the grid. Show each point by writing the capital letter.

a) A (6,1)   b) B (5,7)

c) C (2,6)   d) D (7,4)

e) E (1,3)   f) F (4,2)

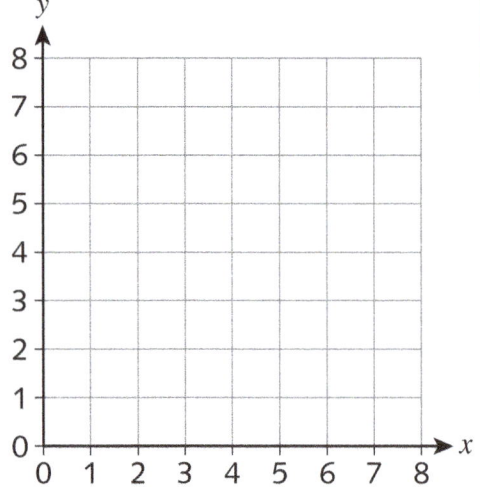

6 marks

96

Geometry – Position and Direction

Coordinates

2   a) A point is plotted at (6,4). It is then moved 1 unit left and 1 unit down.
   What are the new coordinates of the point? _____

b) A point is plotted at (2,1). It is then moved 3 units right and 2 units up.
   What are the new coordinates of the point? _____

2 marks

Marks.......... /8

Challenge 3

1 There are some mystery coordinates.
Work out the coordinates, write them on the lines and plot them on the coordinate grid.

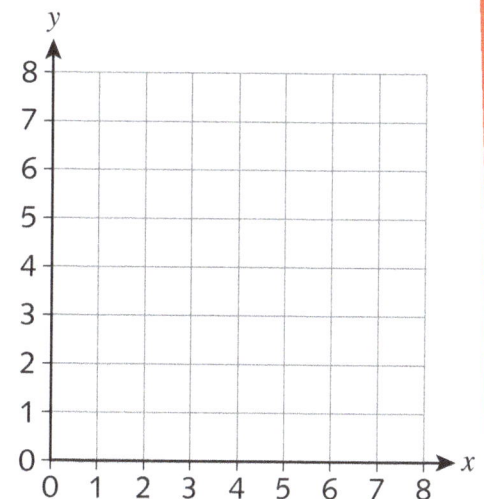

a) The *x*-coordinate and the *y*-coordinate add up to 9.
The cordinates are both multiples of 3.
The *x*-coordinate is the larger number. _____

b) The *x*-coordinate and the *y*-coordinate add up to 10.
The *x*-coordinate is 4 greater than the *y*-coordinate. _____

c) The difference between the coordinates is 2.
The *y*-coordinate is double the *x*-coordinate. _____

d) The *x*-coordinate + the *y*-coordinate = 13.
The *x*-coordinate − the *y*-coordinate = 1. _____

8 marks

Marks.......... /8

Total marks .... /24   How am I doing?

Geometry – Position and Direction

Shapes and Coordinates

PS Problem-solving questions

Challenge 1

1
a) Plot these points on the coordinate grid: (1,7) (1,4) (6,7). They are three vertices of a rectangle.

b) Plot the fourth vertex and use a ruler to complete the rectangle.

c) Write the coordinates of the fourth vertex. _____

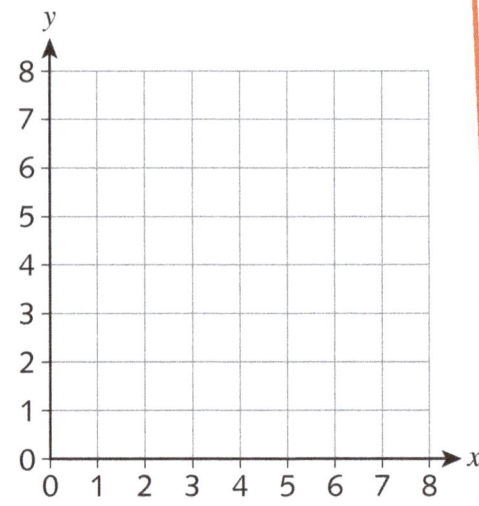

3 marks

2
a) Plot these points on the coordinate grid: (7,3) (3,1). They are two vertices of a right-angled triangle.

b) Plot the third vertex and use a ruler to complete the triangle.

c) Write the coordinates of the third vertex. _____

3 marks

Marks.......... /6

Challenge 2

1
a) Plot these points on the coordinate grid: (3,8) (1,6) (5,6). They are three vertices of a square.

b) Plot the fourth vertex and use a ruler to complete the square.

c) Write the coordinates of the fourth vertex. _____

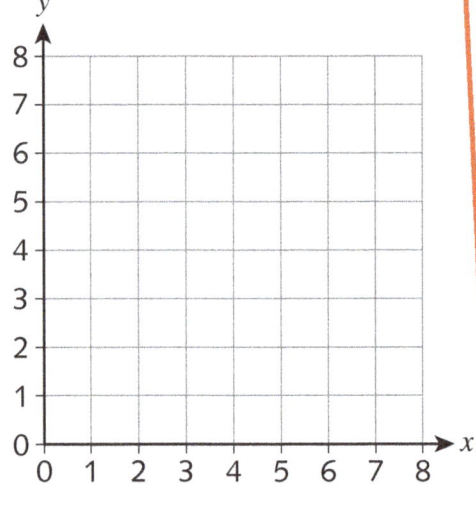

3 marks

2
a) Plot these points on the coordinate grid: (6,6) (4,4) (8,4). They are three vertices of a kite.

b) Plot a possible fourth vertex. Use a ruler to complete the kite.

c) Write possible coordinates of the fourth vertex. _____

3 marks

Marks.......... /6

Geometry – Position and Direction

Shapes and Coordinates

Challenge 3

 1 Three vertices of a square are: (10,15) (10,10) (15,15).

Write the coordinates of the fourth vertex. _____

1 mark

 2 Matt draws a straight line on a coordinate grid.
He uses these coordinates:
(1,5) (3,6) (5,7) (7,8).
Matt extends the line.

Write the next pair of coordinates in the sequence. _____

1 mark

3 a) These coordinates are the end points of a line: (2,2) (8,8).

Write the mid-point of the line. _____

b) These coordinates are the end points of a line: (2,8) (8,2).

Write the mid-point of the line. _____

2 marks

4 Daria says, "If there is a set of coordinates and the first number is always the same, that means the line joining them must be a horizontal line."

Is Daria correct? Explain your answer.

_____

_____

_____

1 mark

Marks.........../5

Total marks ..../17   How am I doing?

Statistics

Bar Charts

PS Problem-solving questions

Look at the bar chart below to answer the questions on pages 100 and 101.
It shows the points scored by five teams on school sports day.

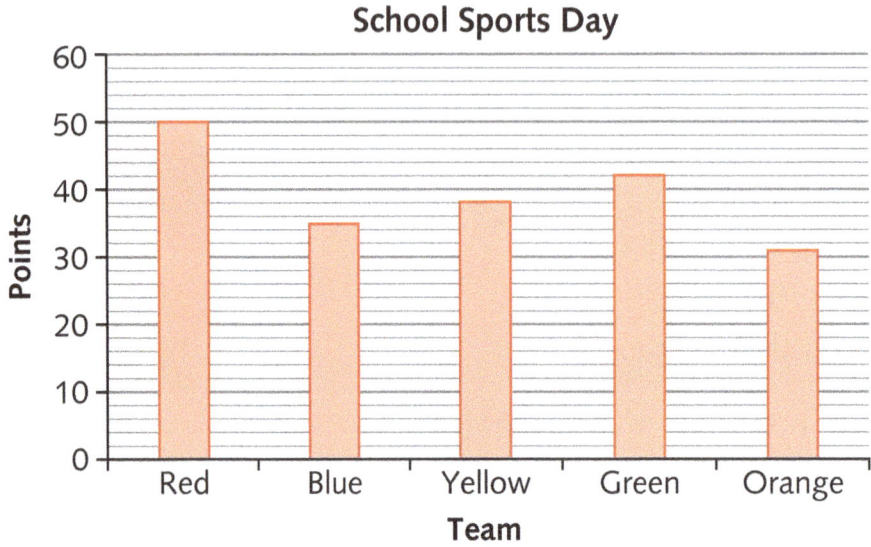

Challenge 1

PS 1
a) What is each small division on the vertical axis worth? _____

b) Which team scored the most number of points? _____

c) Which team scored the least number of points? _____

d) Which team scored 35 points? _____

e) How many more points did the Red team score than the Blue team? _____

5 marks

Marks.......... /5

Challenge 2

PS 1
a) How many teams scored more than 35 points? _____

b) Circle the score for the Yellow team.

Statistics

Bar Charts

c) What was the Green team's score? _____

d) What was the difference in the scores between the Blue and Orange teams? _____

e) What was the difference in the scores between the Yellow and Green teams? _____

5 marks

Marks.......... /5

Challenge 3

1 On sports day, 1st place gained 8 points, 2nd place gained 5 points and 3rd place gained 2 points.

a) What was the points difference between the highest scoring and lowest scoring teams? _____

b) How many points were awarded in total? _____

c) The Green team won three races. Explain how they won the other points.

_____

_____

d) Simon says, "The Red team won Sports Day, but never won a race."

Could Simon be correct? Explain your answer.

_____

_____

_____

4 marks

Marks.......... /4

Total marks .... /14   How am I doing?

Statistics

Time Graphs

PS Problem-solving questions

A class recorded the temperature in the playground throughout one day.
Look at their time graph below to answer the questions on pages 102 and 103.

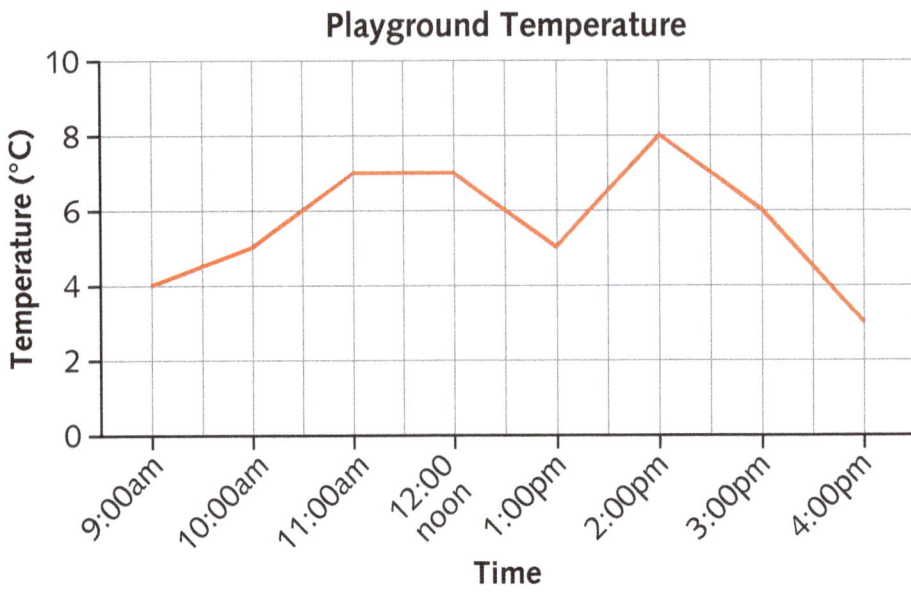

Challenge 1

PS 1 a) Write the time the temperature was first taken.

_____

b) Write the highest temperature.   _____ °C

c) Write the temperature at 3:00pm.   _____ °C

d) For how many hours did the class record the temperature?

_____

e) By how many degrees did the temperature rise between 9:00am and 11:00am?

_____ °C

5 marks

Marks.......... /5

Statistics

Time Graphs

Challenge 2

 1

a) At what time did the temperature first start to go down? _____

b) From the first recording, how long did it take for the temperature to reach 8°C? _____

c) For how long did the temperature stay at 7°C? _____

d) Estimate the temperature at 10:00am. _____ °C

e) Estimate the time the temperature first reached 6°C. _____

5 marks

Marks.......... /5

Challenge 3

PS **1**

a) What is the difference between the temperature at 9:00am and 4:00pm? _____ °C

b) What was the temperature at 12:30pm? _____ °C

c) Estimate the temperature at 9:30am. _____ °C

d) Estimate the time the temperature was above 6°C. _____

e) Estimate the time it took for the temperature to fall from 8°C to 4°C. _____

5 marks

Marks.......... /5

Total marks .... /15   How am I doing?

103

Statistics

Pictograms

PS Problem-solving questions

Challenge 1

PS 1 This pictogram shows the number of rainy days from January to March in four cities.

🌧️ represents 4 rainy days

City	Rainy days
Carlisle	🌧️🌧️🌧️🌧️🌧️🌧️½
Birmingham	🌧️🌧️🌧️🌧️½
Manchester	🌧️🌧️🌧️🌧️🌧️🌧️🌧️🌧️
Norwich	

a) How many rainy days are represented by a rainy cloud symbol? _____

b) How many rainy days were there in Carlisle? _____

c) How many more rainy days were there in Manchester than in Birmingham? _____

d) How many fewer rainy days were there in Carlisle than in Manchester? _____

e) The pictogram is not finished. There were 20 rainy days in Norwich. How many rainy day symbols will complete the pictogram for Norwich? _____

5 marks

Marks.......... /5

Challenge 2

PS 1 This pictogram shows the number of kilometres five friends cycled in one week.

🚲 represents 10 kilometres

Name	Number of kilometres
Sami	🚲🚲🚲🚲🚲🚲🚲🚲🚲🚲🚲🚲
Greg	🚲🚲🚲🚲🚲
Ava	🚲🚲🚲🚲🚲🚲🚲🚲🚲🚲🚲
Martha	🚲🚲🚲🚲🚲🚲🚲🚲🚲🚲
Ricardo	🚲🚲🚲🚲🚲🚲🚲🚲

Statistics

Pictograms

a) Which two friends cycled the same distance?

_____ and _____

b) How far did Martha cycle? _____ km

c) How much further did Sami cycle than Ricardo? _____ km

d) How many friends cycled more than 80 km? _____

4 marks

Marks......... /4

Challenge 3

1 This pictogram shows the number of rooms in different hotels.

	🛏 represents 20 rooms
Hotel	**Number of rooms**
Mill Hotel	🛏 🛏 🛏 🛏 🛏 🛏
Riverbank	🛏 🛏 🛏 🛏
Bridge Hotel	🛏 🛏 🛏 🛏 🛏
Valley Hotel	🛏 🛏 🛏
Old Barn	🛏 🛏

a) How many more rooms are there at the Bridge Hotel than at the Old Barn? _____

b) How many rooms are there at the Old Barn and the Valley Hotel together? _____

c) The Riverbank is only half full. How many rooms are taken? _____

d) 20 rooms at the Mill Hotel are empty. How many rooms are taken? _____

4 marks

Marks......... /4

Total marks .... /13   How am I doing?

Statistics

Tables

PS Problem-solving questions

Challenge 1

PS 1 This table shows the prices of some shirts.

Design \ Size	Small	Medium	Large
Short-sleeved	£12	£14	£16
Long-sleeved	£15	£18	£21

a) What is the cost of a medium long-sleeved shirt? £ _____

b) What is the cost of 2 short-sleeved large shirts? £ _____

c) How much more expensive is a large long-sleeved shirt than a medium short-sleeved shirt? £ _____

d) Gordon buys a small short-sleeved shirt and a small long-sleeved shirt.
How much will he have left from £40? £ _____

4 marks

Marks........../4

Challenge 2

PS 1 This table shows the dinner choices of children in different classes.

Meal \ Class	A	B	C	Total
Hot meal	18	17	12	_____
Vegetarian meal	5	4	_____	18
Sandwich	_____	8	7	20
Total	28	29	28	85

a) How many more children had a hot meal than a vegetarian meal or a sandwich in Class B? _____

106

Statistics

Tables

   b) How many children had a hot meal altogether? Add this to the table.

   c) The number of children who had a sandwich in Class A is missing. Calculate the missing number and add it to the table.

   d) The number of children who had a vegetarian meal in Class C is missing. Calculate the missing number and add it to the table.

4 marks

Marks.......... /4

Challenge 3

PS 1 This table gives information about visitors to a castle.

Visitor \ Day	Friday	Saturday	Sunday	Total
Men	24	18	6	48
Women	___	38	12	76
Children	___	___	16	___
Total	55	71	34	160

   a) How many more men and women visited on Saturday than on Sunday? _____

   b) Olivia says, "As many men visited on Friday as on Saturday and Sunday together." Is Olivia correct? Explain how you know.

   _____

   _____

   c) How many children visited on Friday? _____

   d) Which was the largest group of visitors: men, women or children? _____

4 marks

Marks.......... /4

Total marks .... /12   How am I doing?

Progress Test 4

PS Problem-solving questions

1. Write these numbers in order, starting with the smallest.

   9018   9801   9810   9108   9180

   _____ _____ _____ _____ _____

   1 mark

2. Write the value of the 4 in each number.

   a) 7451 _____   b) 4925 _____   c) 745 _____

   3 marks

3. Write the number of tens altogether in 560. _____

   1 mark

PS 4. Write in the missing numbers.

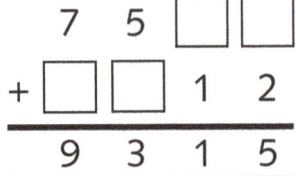

   2 marks

PS 5. Kerry buys some trays of plants for her garden. She buys 14 trays with 4 plants and 15 trays with 6 plants.

   How many plants does she buy altogether? _____ plants

   1 mark

PS 6. A pizza is cut into twelfths. Jonah eats $\frac{3}{12}$ and Ken eats $\frac{4}{12}$.

   How many twelfths are left? _____

   1 mark

PS 7. Larry uses $\frac{1}{5}$ of 5 kg of potatoes on Saturday and $\frac{1}{2}$ of what is left on Sunday.

   How many kilograms are left? _____ kg

   1 mark

8. Write these weights in order, starting with the smallest.

   5500 g $4\frac{1}{2}$ kg   4.9 kg   5 kg   4100 g

   _____ _____ _____ _____ _____

   1 mark

PS 9. Laura spends three weeks of her summer holiday staying with her grandmother.

   How many days is this? _____ days

   1 mark

Progress Test 4

10. Draw lines to match each clock to the equivalent digital clock.

a)  b)  c)

1 mark

11. Convert these measures.

a) 8000 g = _____ kg

b) 9 litres = _____ ml

c) 500 mm = _____ cm

d) 4.5 m = _____ cm

4 marks

 12. Cai has two jugs that each hold 2.25 litres and a third jug that holds 3.75 litres.

How much can the three jugs hold altogether? _____ litres

1 mark

13. Work out:

a) 5 kg − 500 g = _____

b) 5 m − 500 cm = _____

2 marks

14. Below is a triangle. Circle two words, one from each column, to describe the triangle.

acute-angled   **scalene**

right-angled   **isosceles**

obtuse-angled   **equilateral**

2 marks

15. Write the number of lines of symmetry in each shape.

a)  b)

_____ lines of symmetry   _____ lines of symmetry

2 marks

Progress Test 4

PS Problem-solving questions

16. What type of angle is this?

    _____

17. A square is translated from position A to position B. Complete the sentence:

    The square has moved _____ square(s) to the left and _____ square(s) up.

    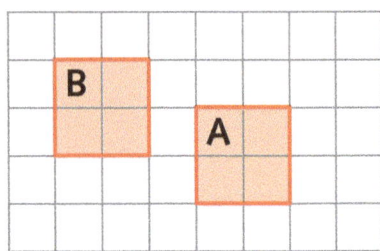

18. Look at the coordinate grid.

    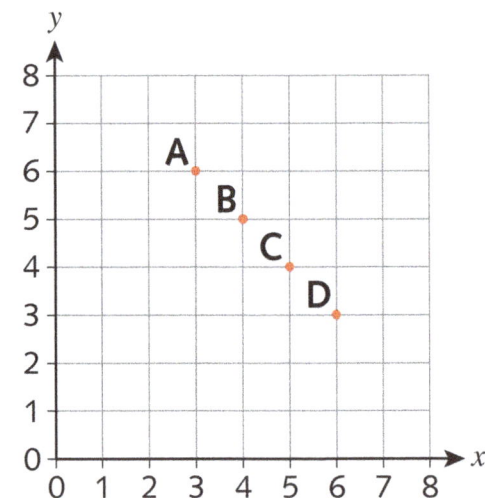

    a) Write the letter at the coordinates (3,6). _____

    b) Carl joins the points A, B, C and D to make a straight line. He makes the line longer; write the coordinates of a point on the line.

    (_____ , _____)

PS 19. This table shows the number of drinks sold in a café one weekend.

Drink \ Day	Teas	Coffees	Milkshakes	Juice	Total
Saturday	45	75	12	23	_____
Sunday	38	62	_____	35	150
Total	83	_____	_____	58	305

a) How many drinks were sold on Saturday? _____ drinks

b) How many coffees were sold altogether? _____ coffees

c) How many milkshakes were sold on Sunday? _____ milkshakes

Progress Test 4

20. This time graph shows the rainfall over one week.

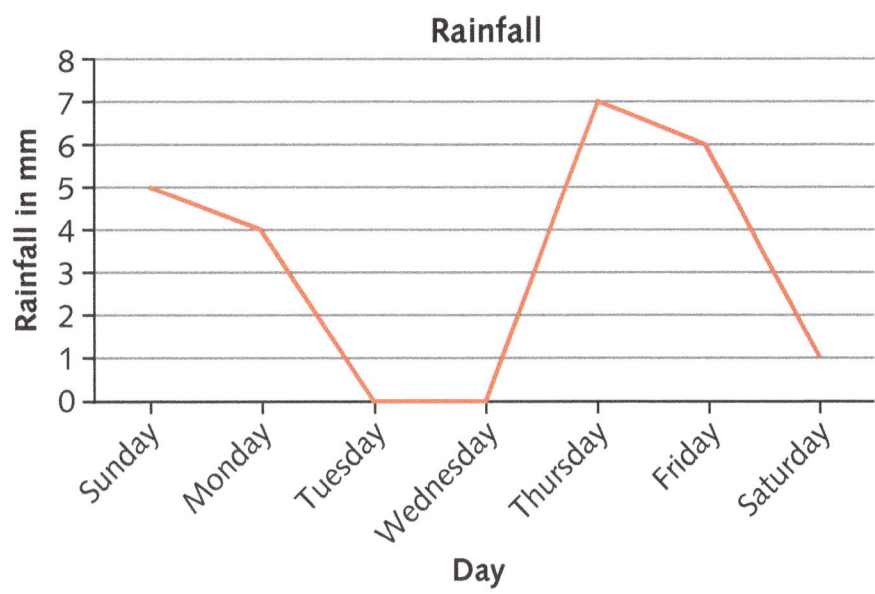

a) What was the greatest amount of rainfall recorded that week? _____ mm

b) Which days had no rain? _____ and _____

c) On how many days was there over 5 mm of rain? _____ days

3 marks

21. This bar chart shows the marks of five children in a test.

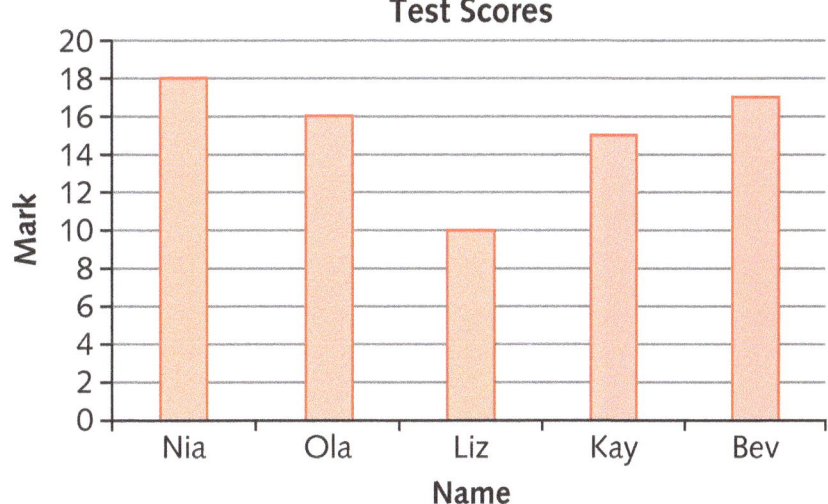

a) Write the name of the person who scored 15. _____

b) Write how many more marks Nia scored than Liz. _____

2 marks

Marks......../36

Notes

Answers

For questions worth one mark with several answer spaces, all answers should be correct to achieve the mark, unless otherwise indicated.

Pages 4–11
Starter Test
1. a) 32, 36   b) 550, 600
2. 760
3. a) <   b) <
4. 871 circled
5. a) 25   b) 70
6. a) 821   b) 576   c) 358
7. 252
8. 635
9. a) 443   b) 384
10. 693 − 265
11. a) 54   b) 96   c) 112
12. a) 14   b) 19   c) 18
13. 72
14. 23
15. 48p
16. a) $\frac{1}{2}$ (accept equivalent fractions)
    b) $\frac{7}{8}$
17. $\frac{1}{10}$
18. a) 4   b) 16   c) $\frac{4}{5}$
19. a) $\frac{2}{5}$ (accept $\frac{4}{10}$ or 0.4)
    b) $\frac{9}{10}$ (accept 0.9)
20. a) $\frac{2}{6}$ b) $\frac{4}{8}$
21. a) <   b) >
22. a) $\frac{2}{5}$ b) $\frac{3}{10}$ c) $\frac{7}{8}$
23. one
24. $\frac{7}{8}$
25. 1 mm, 1 cm, 1 m
26. 2 m circled
27. a) >   b) >
28. 0.5 kg ticked
29. a) >   b) >
30. 4 litres
31. 3 kg (accept any weight ≥2.8 kg but ≤3.2 kg)
32. a) 400 ml (accept any capacity ≥380 ml but ≤420 ml)
    b) 250 ml (accept any capacity ≥230 ml but ≤270 ml)
33. a) 2:20pm   b) 9:50am
    c) 8:30am   d) 9:45pm
34. a) 9:15pm   b) 6:55am
35. 180 seconds
36. 30th September circled
37. 365 days
38. 50 minutes
39. 16 cm
40. £12.30
41. £89.85
42. a) 800 cm   b) 11 kg
    c) 2 litres   d) 45 mm
43. Line **B** ticked
44. Lines circled in Hexagon **C**
45. A pentagon with one right angle drawn
46. a) cuboid   b) cone   c) 6
    d) circle
47. Angles **A** and **C** ticked
48. 2
49. a) 55   b) 64
50. a) Ben   b) 3   c) 5
    d) Bar for Gus, representing 5 lengths, added to chart
51. a) 8   b) 4   c) 6

Pages 12–13
Challenge 1
1. a) 40, 48   b) 48, 54
2. a) Add 6   b) Add 25
3. No
4. 56

Challenge 2
1. a) 42, 48   b) 200, 225
2. a) Add 7   b) Add 25
3. 650

Challenge 3
1. 135, 144, 153
2. 60, 66, 72
3. 71
4.

	Add 7 →			
21	28	35	42	49
30	37	44	**51**	58
39	**46**	53	60	67
48	55	62	69	76

(Add 9 ↓)

Pages 14–15
Challenge 1
1. a) <   b) >   c) <
2. Statement **C** ticked
3. 3465, 3563, 4536, 4645, 4654
4. 7541

Challenge 2
1. a) >   b) <   c) <
2. 4138, 4831, 5246, 5264, 5642
3. a) 865   b) 1356

113

Answers

Challenge 3
1. 681, 5891, 7861, 11 937, 12 612
2. Any number >6285 but <6290
3. Athletic vs Wanderers

Pages 16–17
Challenge 1
1. a) 40   b) 400   c) 4
2. 1475 circled
3. 341 circled
4. 3627

Challenge 2
1. a) 900   b) 90   c) 9000
2. 3890
3. 9738
4. 6342
5. a) Any combination of four digits with 5 or 6 as the first digit.
   b) 2345

Challenge 3
1. Jamie is incorrect because 4021 has four digits (thousands), which is more than 897 which only has three digits (hundreds).
2. 6031
3. a) 10 009   b) 10 090

Pages 18–19
Challenge 1
1. a) 12   b) 15   c) 26
   d) 8   e) 33   f) 40
2. a) 3 + 2 = 5   b) 6 + 4 = 10
   c) 26 – 7 = 19
3. a) 4:40 **or** twenty to five
   b) 10:20 **or** twenty past ten

Challenge 2
1. a) 56   b) 48   c) 96   d) 83
2. 90
3. a) 40 + 60 = 100   b) 69 – 30 = 39
4. XCIX circled

Challenge 3
1. XXXVIII, XLVII, LXV, LXIX, XC
2. a) 46   b) 136   c) 50   d) 84
3. a) 125   b) 222   c) 359

Pages 20–21
Challenge 1
1. a) 200   b) 400
   c) Any number >420 but <480
2. a) 351   b) 524
3. a) 5296   b) 7570   c) 2804

Challenge 2
1. 457
2. a) 4000 + 800 + 20 + 5 (accept alternatives such as 3000 + 1800 + 20 + 5)
   b) 6000 + 700 + 10 + 9 (accept alternatives such as 3000 + 3700 + 10 + 9)
3. a) Any number >3450 but <3550
   b) Any number >4100 but <4400
   c) Any number >4600 but <4750

Challenge 3
1. a)–e) Accept any five different additions that total 3746
2. a) 4609   b) 8020
3. a) 7 thousands = **70** hundreds = **700** tens = **7000** ones
   b) 800   c) 3000
4. 2738

Pages 22–23
Challenge 1
1. a) 50   b) 100   c) 80
2. 84, 75 and 77 circled
3. 120
4. a) 300   b) 700   c) 1500

Challenge 2
1. a) 3800   b) 6300   c) 5100
2. a) 2630   b) 2600   c) 3000
3. a) 4000   b) 3000   c) 8000
4. Any three numbers ≥650 but <750

Challenge 3
1. a) 6000   b) 1000   c) 4000
2. Any number ≥7335 but <7345
3. 350, 6000 and 7900 circled
4. 6951, 6915, 6591, 6519

Pages 24–25
Challenge 1
1. a) –2   b) –4   c) –2   d) –5
2. a) –2   b) –4   c) –1
3. £4
4. –5

Challenge 2
1. a) –4   b) –7   c) –2   d) –9
2. a) –5   b) –5   c) –1
3. a) –9   b) –17   c) –18
4. a) –2°C   b) –9°C   c) 4°C

Challenge 3
1. 5, 3, –1, –4, –6
2. a) –6   b) –4   c) 0   d) 4
3. a) –9   b) –15   c) –8
   d) –14   e) –24   f) –4

Pages 26–27
Challenge 1
1. a) 749   b) 997   c) 1005
2. a) 401   b) 91   c) 528

114

Answers

Challenge 2
1. a) 1528 b) 1418 c) 13728
2. a) 643 b) 406 c) 3596

Challenge 3
1. a) 12400 b) 11312 c) 12909
2. a) 1196 b) 3973 c) 4793

Pages 28–29
Challenge 1
1. a) 120 b) 90 c) 170
2. 390 – 265 circled
3. a) 563 – 328 or 563 – 235
   b) 293 + 236 c) 311 + 214
4. 227

Challenge 2
1. 356
2. 630
3. a) 1000 b) 100
4. a) 3000 b) 11000

Challenge 3
1. Possible answers are: 84p + 51p, 83p + 52p, 82p + 53p, 81p + 54p
2. a) 17000 b) 5000
3. 9258 – 3826 = 5432
   Correct answer = 10258
4. 347

Pages 30–31
Challenge 1
1. a) 147 b) 425
2. 613
3. £54
4. 125 cm

Challenge 2
1. 6797
2. 3303
3. 2093 km
4. 2181

Challenge 3
1. a) 3874 + 2643 b) 8374 – 2650
2. Accept different possible answers: 5 and 3 are hundreds, 1 and 2 are tens, and 6 and 4 are ones, e.g. 526 + 314
3. A 5236 + 1609 = **6845**
   B 9343 – 2434 = **6909**
   C 2354 + 4567 = **6921** circled
   (1 mark for each calculation worked out correctly; 1 mark for circling calculation C)

Pages 32–33
Challenge 1
1. a) 35 b) 24 c) 88
   d) 36 e) 70 f) 32
2. a) 7 b) 12 c) 7
   d) 6 e) 10 f) 6
3. a) 8 b) 9 c) 9
   d) 7 e) 6 f) 12

Challenge 2
1. a) 42 b) 28 c) 99
   d) 56 e) 81 f) 84
2. a) 7 b) 8 c) 12
   d) 12 e) 12 f) 9
3. a) 11 b) 3 c) 11
   d) 12 e) 6 f) 4

Challenge 3
1. a)

×	7	9	12
7	49	**63**	**84**
9	**63**	**81**	**108**
12	**84**	**108**	**144**

b)

×	7	8	9
6	42	48	54
8	56	64	**72**
12	84	96	108

(1 mark for each correct row)

2. a) Possible answers are: 48 × 1, 24 × 2, 16 × 3, 12 × 4
   b) Possible answers are: 36 × 1, 18 × 2, 12 × 3, 6 × 6
   c) Possible answers are: 24 × 1, 12 × 2, 6 × 4
   d) Possible answers are: 72 × 1, 36 × 2, 24 × 3, 18 × 4, 9 × 8
3. a) 81 b) 56 c) 64 d) 120

Pages 34–35
Challenge 1
1. a) 5 b) 0 c) 0
   d) 60 e) 160 f) 350
2. a) 20 b) 20 c) 30
3. a) 40 b) 30 c) 50
4. a) 3 b) 5 c) 4

Challenge 2
1. a) 30 b) 110 c) 64
   d) 69 e) 128 f) 255
2. a) 15 b) 13 c) 14
   d) 34 e) 16 f) 22
3. a) 48 b) 90 c) 72

Answers

Challenge 3
1. a) 3 b) 4 c) 5
   d) 3 e) 5 f) 3
2. a) 3 b) 6 c) 4
   d) 4 e) 12 f) 3
3. a) 4 b) 3 c) 15 d) 24

Pages 36–37
Challenge 1
1. A, D and F circled
2. a) 5 b) 12
3. 9
4. 1, 3, 5, 10 and 15 circled

Challenge 2
1. a) 1, 2, 5, 10
   b) 1, 3, 7, 21
   c) 1, 2, 4, 8, 16, 32
   d) 1, 2, 4, 5, 8, 10, 20, 40
2. 2 and 21
3. 7 and 8
4. Statement B circled
5. 7

Challenge 3
1.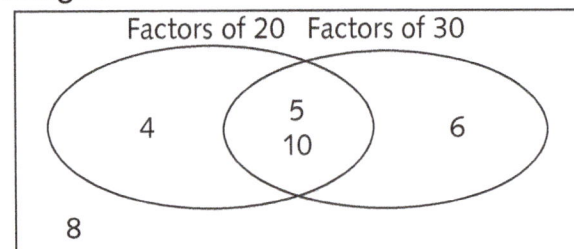
2. 6 rows of 8 bottles
3. a) 48 × 1, 24 × 2, 16 × 3, 12 × 4, 8 × 6
   b) 60 × 1, 30 × 2, 20 × 3, 15 × 4, 12 × 5, 10 × 6
   c) 1 × 100, 2 × 50, 4 × 25, 5 × 20, 10 × 10
4. 8, 56

Pages 38–39
Challenge 1
1. a) 230 b) 318 c) 248 d) 111
2. a) 216 (calculation set out as a column multiplication)
   b) 360 (calculation set out as a column multiplication)
   c) 672 (calculation set out as a column multiplication)
3. a) 4 b) 3

Challenge 2
1. a) 1518 b) 3035 c) 4459
2. a) 725 × 6 = 700 × 6 + 20 × 6 + 5 × 6
      = 4350
   b) 2

3. a) 2989 (calculation set out as a column multiplication)
   b) 4098 (calculation set out as a column multiplication)
   c) 3843 (calculation set out as a column multiplication)

Challenge 3
1. a) 5992 (calculation set out as a column multiplication)
   b) 7736 (calculation set out as a column multiplication)
   c) 8604 (calculation set out as a column multiplication)
2. a) 547 × 7 = 3829 b) 385 × 9 = 3465
   c) 276 × 8 = 2208
3. 643 × 6 = 3858

Pages 40–41
Challenge 1
1. 288
2. 280
3. 60
4. £8.75
5. 168

Challenge 2
1. 744
2. 1536
3. £4170
4. 1390
5. 3800

Challenge 3
1. a) £2430 b) £2928 c) £2790
   d) £5176 e) £2422

Pages 42–45
Progress Test 1
1. a) 72, 78 b) 875, 900
2. 5274, 5427, 5472, 5724, 5742
3. a) < b) <
4. £7799
5. a) 5835 b) 5008
6. 2390 circled
7. 8354
8. a) 38 b) 64
9. XLII circled
10. 535
11. a) 40
    b) Any number ≥184 but ≤186
12. a) 500 b) 800 c) 100

Answers

13.

	Rounded to the nearest 10 is	Rounded to the nearest 100 is	Rounded to the nearest 1000 is
5826	5830	5800	6000

14. −10, −20
15. −5
16. a) 240 b) 1080 c) 545
17. a) 527 − **353** = 174   b) **358** + 551 = 909
18. 372 + 322 circled
19. 282 (accept 365)
20. 599 minutes
21. 33 kilometres
22. a) 35 b) 27 c) 84
 d) 8 e) 12 f) 6
23.

×	6	9	12
4	24	36	48
9	54	81	**108**
7	42	63	84

(1 mark for each correct row)

24. a) 75 b) 16 c) 88
 d) 18 e) 12 f) 102
25. a) 1 and 20, 2 and 10, 4 and 5
 b) 1 and 100, 2 and 50, 4 and 25, 5 and 20, 10 and 10
26. a) 378 b) 1060 c) 4256
27. 24
28. 54 × 3 = 50 × 3 + **4** × **3** = 162 (accept 4 and 3 in either order)
29. 35 metres
30. 100

Pages 46–47
Challenge 1
1. a) $\frac{3}{10}$ b) $\frac{9}{10}$ c) $\frac{7}{10}$
2. a) $\frac{9}{100}, \frac{13}{100}$ b) $\frac{79}{100}, \frac{80}{100}$ (accept $\frac{8}{10}$), $\frac{81}{100}$
3. 100

Challenge 2
1. a) $\frac{7}{100}$ b) $\frac{43}{100}$ c) $\frac{89}{100}$
2. a) $\frac{6}{100}$ (accept $\frac{3}{50}$) b) $\frac{9}{100}$
 c) $\frac{2}{100}$ (accept $\frac{1}{50}$)
3. $\frac{1}{20}$ of a litre (accept $\frac{5}{100}$ or equivalent)

4. 1 cm

Challenge 3
1. 5 ÷ 100 = $\frac{5}{100}$ circled
2. 1 ÷ 100 and 0.1 ÷ 10 circled
3. a) 7 b) 5 c) 0.7
4. a) 6 b) 10

Pages 48–49
Challenge 1
1. a) $\frac{3}{6} = \frac{1}{2}$ b) $\frac{4}{8} = \frac{1}{2}$
2. a) $\frac{5}{10}$ b) $\frac{6}{10}$ c) $\frac{8}{10}$
3. a) $\frac{4}{8}$ b) $\frac{2}{6}$ c) $\frac{3}{4}$

Challenge 2
1. a) $\frac{10}{20}$ b) $\frac{6}{20}$ c) $\frac{12}{20}$
2. a) $\frac{5}{15}$ b) $\frac{10}{15}$ c) $\frac{9}{15}$
3. a) $\frac{4}{12}, \frac{5}{15}, \frac{6}{18}, \frac{7}{21}, \frac{8}{24}$ b) $\frac{12}{16}, \frac{15}{20}, \frac{18}{24}, \frac{21}{28}, \frac{24}{32}$

Challenge 3
1. a) $\frac{10}{20}$ b) $\frac{16}{20}$ c) $\frac{6}{20}$
 d) $\frac{1}{4}$ e) $\frac{2}{5}$ f) $\frac{9}{10}$
2. $\frac{8}{30}$ and $\frac{20}{100}$ circled
3. $\frac{2}{5}$

Pages 50–51
Challenge 1
1. a) $\frac{3}{5} + \frac{1}{5} = \frac{4}{5}$ b) $\frac{5}{6} - \frac{4}{6} = \frac{1}{6}$
2. a) $\frac{3}{3}$ or 1 b) $\frac{13}{8}$ or $1\frac{5}{8}$
 c) $\frac{17}{10}$ or $1\frac{7}{10}$
3. a) $\frac{3}{5}$ b) $\frac{3}{10}$ c) $\frac{1}{8}$

Challenge 2
1. a) $\frac{4}{5} + \frac{2}{5} = \frac{6}{5} = 1\frac{1}{5}$
 b) $\frac{4}{6} + \frac{4}{6} = \frac{8}{6} = 1\frac{2}{6}$ or $1\frac{1}{3}$
2. a) $\frac{4}{3} = 1\frac{1}{3}$ b) $\frac{7}{5} = 1\frac{2}{5}$
 c) $\frac{6}{4} = 1\frac{2}{4}$ or $1\frac{1}{2}$
3. a) $\frac{3}{10}$ b) $\frac{5}{12}$ c) $\frac{9}{10}$

Challenge 3
1. a) $\frac{5}{5} - \frac{3}{5} = \frac{2}{5}$ (accept $1 - \frac{3}{5} = \frac{2}{5}$)
 b) $\frac{3}{4} + \frac{3}{4} = \frac{6}{4}$ or $1\frac{2}{4}$ or $1\frac{1}{2}$

Answers

2. a) $\frac{5}{10}$ (accept $\frac{1}{2}$)   b) $\frac{6}{12}$ (accept $\frac{1}{2}$)
   c) $\frac{9}{8}$ (accept $1\frac{1}{8}$)
3. a) $\frac{5}{8}$ b) $\frac{6}{7}$ c) $\frac{3}{10}$

Pages 52–53
Challenge 1
1. a) 5   b) 6
   c) 6 m   d) 4 kg
2. a) 16   b) 16
   c) 27 ml   d) 25 km
3. 2 km
4. 14

Challenge 2
1. a) 7   b) 4
   c) £2   d) 2 m
2. a) 14   b) 28
   c) 55 cm   d) 6 km
3. 24 kg
4. 30

Challenge 3
1. a) 35   b) 45
2. 9
3. a) 20   b) 30
4. 15
5. Explanation that shows half of 27 is not a whole number or that $13\frac{1}{2}$ boys or girls cannot be possible.
6. a) 4   b) 20

Pages 54–55
Challenge 1
1. $\frac{3}{4}$ = 0.75, $\frac{1}{2}$ = 0.5, $\frac{1}{4}$ = 0.25
2. a) 0.1   b) 0.7   c) 0.3
3. a) 0.9   b) 1.6   c) 2.4   d) 5.5
4. a) 0.4   b) 0.8   c) 1

Challenge 2
1. a) 0.5   b) 0.25   c) 0.75
2. a) 0.9   b) 0.57   c) 0.03
3. £4.57
4. a) 0.3   b) 0.65   c) 0.95

Challenge 3
1. a) 0.09   b) 0.2   c) 1.01
2. a) 0.15
   b) Any number >0.45, <0.48
   c) Any number >0.8, <0.84
3. James is incorrect because 0.7 as a fraction is $\frac{7}{10}$, whereas $\frac{7}{100}$ is 0.07.
4. a) 10   b) 10

Pages 56–57
Challenge 1
1. a) 5   b) 5   c) 3
   d) 6   e) 8   f) 3
2. 3 m
3. 1 litre
4. Any number ≥6.5 but <7

Challenge 2
1. a) 7   b) 7   c) 22
   d) 54   e) 86   f) 105
2. 50 kg
3. Any number ≥49.5 but <50.5
4. 29.7 and 30.4 circled

Challenge 3
1. a) 116   b) 230   c) 707
2. 100.3 and 99.6 circled
3. 13
4. 48.7 and 49.4 → 49
   49.8, 50.4 and 49.5 → 50

Pages 58–59
Challenge 1
1. 0.8 circled
2. 2.4 circled
3. 4.7, 5.6, 5.9, 6.2
4. Craig
5. Any number >7.6 but <8

Challenge 2
1. 12.16 circled
2. 34.45, 35.44, 43.54, 45.34, 45.43
3. 60.56
4. Any number >23.56 but <23.62
5. Any number >45.34 but <45.43 in the first space and any number >45.27 but <45.34 in the second space

Challenge 3
1. a) <   b) >
2. a) 3.76 m   b) 2.88 m
3.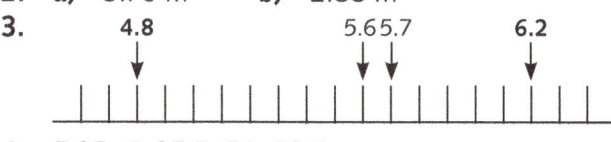
4. 5.25, 6, 25.5, 50, 52.5

Pages 60–61
Challenge 1
1. a) 0.6   b) 1.2   c) 0.07
   d) 0.25
2. a) 5   b) 36   c) 6.3
3. a) 0.7 (accept seven tenths)
   b) 7 (accept seven ones)
   c) 0.07 (accept seven hundredths)
4. 0.25 kg

Answers

Challenge 2
1. a) 0.67 b) 0.54
   c) 0.15 d) 7.8
2. a) 5.7 b) 36
   c) 100 d) 100
3. 60
4. a) 0.04 (accept four hundredths)
   b) 0.04 (accept four hundredths)
5. £0.64 (accept 64p if £ sign is crossed out)

Challenge 3
1. a) 10 b) 100
   c) 32 d) 79
2. Many possible solutions:
   Kagendo's calculation, e.g. 9 ÷ 100 = 0.09
   Sally's calculation, e.g. 9 ÷ 10 = 0.9
3. 6.7 ÷ 10 and 56 ÷ 10 circled
4. a) 45 ÷ 10
   b) 60.8 ÷ 10

Pages 62–63
Challenge 1
1. a) £9.25 b) £5.05
   c) A set of pens and a book
2. 3 litres
3. 25 km
4. £2.40

Challenge 2
1. £5.11
2. £1.05
3. 1.35 kg
4. £8.10

Challenge 3
1. 12 kg
2. 10
3. a) 22.2 km b) 20.8 km c) 15.4 km

Pages 64–67
Progress Test 2
1. a) 77, 84
   b) 7000, 8000
2. 1384, 1834, 4138, 8314, 8341
3. 1 ten circled
4. a) and b) Accept any two different combinations of numbers that total 538, e.g. 300 + 200 + 38
5.

Number	Rounded to the nearest 10
3784	**3780**
427	**430**
79	**80**

6. −6
7. a) 6998 b) 4746
8. 600 + 400 = 1000
9. 318
10. 149 m
11. a) 72 b) 240
12. 168
13. a) 2128 b) 7520 c) 3768
14. 4 children share **6** paint pots and **10** brushes.
15. 48 × 3 = **40** × 3 + **8** × 3 = 144
16. a) $\frac{3}{100}$ (accept 0.03)
    b) $\frac{9}{100}$ (accept 0.09)
    c) $\frac{7}{100}$ (accept 0.07)
17. a) $\frac{7}{100}$ b) $\frac{16}{100}$
18. a) $\frac{3}{4}$ b) $\frac{5}{10}$
19. a) $\frac{11}{10}$ or $1\frac{1}{10}$ b) $\frac{9}{8}$ or $1\frac{1}{8}$
20. a) $\frac{13}{100}$ b) $\frac{9}{10}$ (accept $\frac{90}{100}$) c) $\frac{9}{100}$
21. a) 0.93 b) 0.07 c) 0.4
22. a) 5 b) 15 c) 30
23. Accept any of the following: 24.5, 24.6, 24.7, 24.8, 24.9
24. 73.8 circled
25. 30.2, 30.7, 31.6, 31.8, 32.5
26. a) 10 b) 100 c) 100 d) 10
27. 57
28. a) 10 b) 24 c) 81
29. £29 (accept £29.00; do not accept £29.0)

Pages 68–69
Challenge 1
1. 83.76 m, 83.67 m, 63.78 m, 38.76 m, 36.87 m
2. 4.87 litres, 5.47 litres, 45.78 litres, 54.87 litres, 78.54 litres
3. Accept any capacity >4.56 litres but <4.75 litres
4. 1.25 m, $1\frac{1}{2}$ m, 1.65 m
5. Dubry

Challenge 2
1. a) 3.2 kg circled b) 8000 g circled
2. a) 474 cm circled b) 6 cm circled
3. a) 5.675 litres circled b) 10 litres circled
4. 4.7 m, 475 cm, 4800 mm, 5 m
5. No

Challenge 3
1. 4.8 m circled
2. 4500 metres circled
3. 1 litre bottle circled
4. B, C, A

Answers

Pages 70–71
Challenge 1
1. Accept any mass >3.5 kg but <3.9 kg
2. Accept any length >7.4 cm but <7.8 cm
3. 0.75 litre
4. £24.75 and £26.15 ticked

Challenge 2
1. 4 m ticked
2. 350 ml ticked
3. 120 g ticked
4. 12 minutes ticked

Challenge 3
1. £15
2. Accept any temperature >26°C but <30°C
3. Accept any length >7 cm but <10 cm
4. Accept any mass >2 kg but <2.5 kg

Pages 72–73
Challenge 1
1. a) 3 cm   b) 4 m   c) 5 km
   d) 60 mm   e) 500 cm   f) 9000 m
   g) 4 kg   h) 40 kg   i) 5000 g
   j) 8000 ml   k) 10 litres   l) 500 ml
2. 4
3. a) Yes   b) 5 cm or 50 mm

Challenge 2
1. a) 520 cm   b) 3500 g   c) 3.5 litres
   d) 7.8 cm   e) 0.25 km   f) 75 cm
   g) 8.45 m   h) 10.5 km
2. 400 g or 0.4 kg
3. Jug B
4. 68 ml

Challenge 3
1. 1000 mm
2. a) m   b) cm   c) km
   d) litres   e) g   f) cm
3.

| $\frac{1}{2}$ litre | 300 ml | 200 ml |

| | | 0.6 litres |

| $\frac{1}{4}$ litre | 200 ml | 550 ml |

(accept equivalent measures, e.g. $\frac{1}{5}$ litre)

Pages 74–75
Challenge 1
1. £1.55
2. £13.24
3. £4 (accept £4.00; do not accept £4.0)
4. £4.95
5. £2.67

Challenge 2
1. a) £7.10   b) 3.76 m
   c) 18.20 kg   d) 46.79 litres
2. £36.60
3. 6.5 kg
4. 9.25 m
5. 15 litres

Challenge 3
1. a) 3.55 kg   b) 3550 g
2. 50p
3. Accept three 0.6 m cupboards and three 0.4 m cupboards **or** six 0.4 m cupboards and one 0.6 m cupboard.
4. 4

Pages 76–77
Challenge 1
1. a) 4:10am   b) 7:25am   c) 4:55am
2. a) 18:45   b) 17:05   c) 21:35
3. a) Twenty past eight **or** 20 past 8
   b) Five past eight **or** 5 past 8
   c) Twenty to eleven **or** 20 to 11

Challenge 2
1. a) 3:25pm   b) 6:10pm
2. a) 14:15   b) 11:50   c) 18:35
3. 12:15 **or** quarter past 12 **or** $\frac{1}{4}$ past 12 **or** 00:15

Challenge 3
1.

	Word time	12 hour time		24 hour time
a)	Five past two in the afternoon	= 2:05pm	=	14:05
b)	Twenty-five to 9 in the evening	= 8:35pm	=	20:35

2. 40 minutes

Pages 78–79
Challenge 1
1. a) 120 seconds   b) 120 minutes
   c) 48 hours   d) 14 days
2. April, June, September and November
3. 2024
4. 8th May

Challenge 2
1. Accept 2 hours 20 minutes **or** 140 minutes
2. 40 minutes
3. Accept 19:15, 7:15, 7:15pm **or** quarter past 7
4. a) 28 days   b) 24 months
   c) 300 seconds   d) 72 hours

Challenge 3
1. Accept any time >8 minutes but <10 minutes
2. a) **336** hours = **14** days = 2 weeks
   b) **120** months = **10** years = 1 decade
3. Accept 06:45, 6:45, 6:45am **or** quarter to 7

Answers

4. 3 hours 5 minutes
5. Accept 13:27, 1:27, 1:27pm **or** 27 past 1

Pages 80–81
Challenge 1
1. A a) 8 squares b) 12 cm
   B a) 9 squares b) 12 cm
   C a) 10 squares b) 22 cm

Challenge 2
1. A a) 9 squares b) 16 cm
   B a) 16 squares b) 20 cm
   C a) 7 squares b) 16 cm
2. 28 cm

Challenge 3
1. a) 2 cm b) 5 small squares
2. 14 cm

Pages 82–85
Progress Test 3
1. a) 63, **72**, 81, **90**, 99, 108
   b) 36, **42**, **48**, 54, 60, 66
2. 3752, 3727, 3725, 3572, 3527
3. 6548
4. LXV circled
5.

Number	Rounded to the nearest hundred
7 8 2 3	7800
7 8 2	800
7 8	100

6. a) 545 b) 502
7. a) 6 b) 6
   c) 5 d) 49
8. 8
9. a) 4581 b) 3164 c) 5166
10. 6 hours
11. a) 100 b) 10 c) 10
12. A line drawn from $\frac{3}{5}$ to $\frac{6}{10}$
13. a) $\frac{1}{8}$ b) $\frac{3}{12}$
14. £19
15. £10
16. a) Balloons, candles and a cake
    b) £2.55
17. £3, £1.25, 90p, £0.08, 5p
18. a) Tia b) Raf
19. £40
20. 15 minutes
21. Clock **B** circled
22. Statement **A** ticked
23. 1 minute 40 seconds

24. a) 5000 g b) 6 cm c) 10 m
    d) 8 litres e) 800 cm f) 2000 m
25. a) 13 cm b) 12 cm
26. 25 squares

Pages 86–87
Challenge 1
1. Shapes **A** and **C** ticked
2. a) **B** and **D** b) **A** and **D**
   c) **C** d) **B**
3. Shape **B** ticked

Challenge 2
1. Shapes **B** and **D** ticked
2. A hexagon has six sides. This is a pentagon as it has five sides.
3. Triangle **B** ticked

Challenge 3
1. Octagon circled
2. Shapes **A**, **B** and **C** ticked
3. Any pentagon drawn with a right angle.

Pages 88–89
Challenge 1
1. Shape **A** ticked
2. Shape **C** ticked
3. a) 6 b) 6

Challenge 2
1. a) 7 b) 4
2. Statement **D** ticked
3.

Shape	Faces	Edges	Vertices
Cuboid	6	12	8
Square-based pyramid	5	8	5
Triangular prism	5	9	6

(1 mark for each correct row)

Challenge 3
1. Yes, Shona is correct. On each face, the angles at the vertices are right angles and the opposite edges are equal and parallel. These are the properties of a rectangle.
2. Pentagon
3. a) 5 b) 7

Pages 90–91
Challenge 1
1.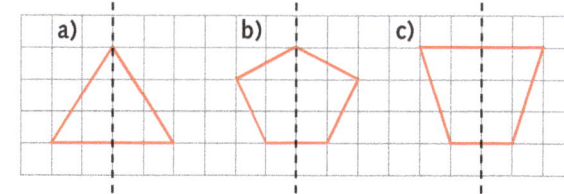

121

Answers

2.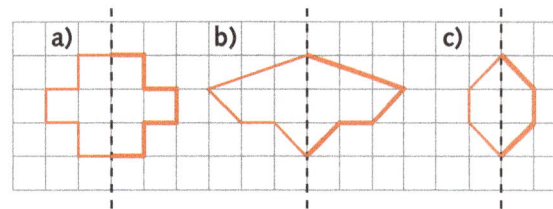

Challenge 2
1. a) 4   b) 2   c) 2
2.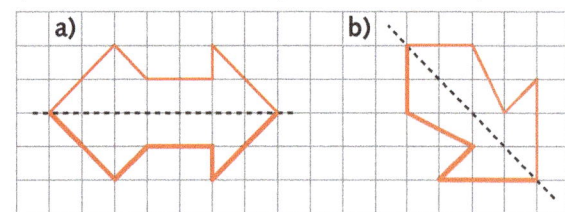

Challenge 3
1. Shape **D** ticked
2. Different answers are possible, e.g.:

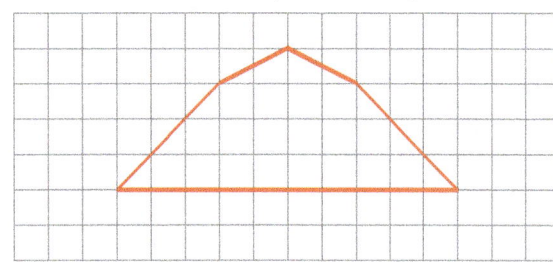

or

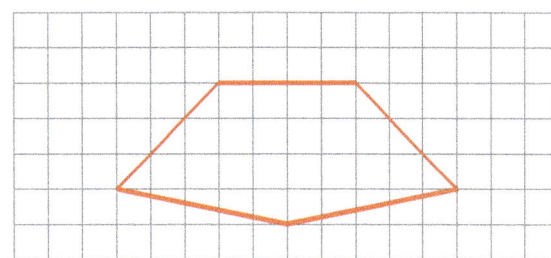

Pages 92–93
Challenge 1
1. a) 4   b) 6   c) 4
2. Angle **B** ticked
3. Angle **A** ticked
4. Angle **D** ticked

Challenge 2
1.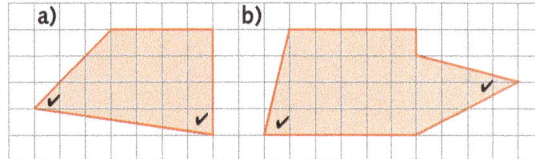

2. B, D, A, C
3. obtuse, right, acute

Challenge 3
1. a) Any five-sided shape with three obtuse angles drawn, e.g.

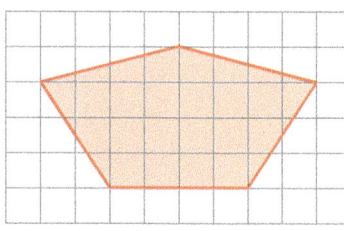

b) Any four-sided shape with three acute angles drawn, e.g.

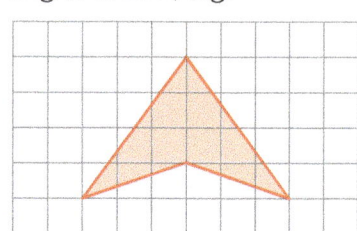

2. No, Caroline isn't correct. Triangles cannot have two right angles.

Pages 94–95
Challenge 1
1. a) C   b) D   c) F
   d) A   e) B
   f) Square **B** to square **D**

Challenge 2
1. a) 11 right, 1 down   b) 3 left, 3 up
   c) 7 left, 1 down   d) 11 right, 4 up
   e) 4 right, 2 down   f) 4 left, 3 down

Challenge 3
1. 5 (units) right, 7 (units) up
2.
3. a) row circled   b) column circled

Pages 96–97
Challenge 1
1. a) (4,6)   b) (2,7)   c) (6,4)
   d) (5,3)   e) (7,2)   f) (3,5)
2. a) Points (1,2) and (7,8) plotted and joined with a straight line.
   b) (4,5)

Challenge 2
1. a) A plotted at (6,1)   b) B plotted at (5,7)
   c) C plotted at (2,6)   d) D plotted at (7,4)
   e) E plotted at (1,3)   f) F plotted at (4,2)

Answers

2. **a)** (5,3)   **b)** (5,3)

Challenge 3
1. **a)** (6,3) written and plotted on the grid
   b) (7,3) written and plotted on the grid
   c) (2,4) written and plotted on the grid
   d) (7,6) written and plotted on the grid
   (1 mark for each pair of coordinates written correctly; 1 mark for each pair of coordinates plotted correctly)

Pages 98–99
Challenge 1
1. **a)** and **b)** Rectangle plotted with vertices at (1,7), (1,4), (6,7) and (6,4)
   c) (6,4)
2. **a)** (7,3) and (3,1) plotted
   b) A right-angled triangle drawn
   c) The third vertex could be: (0,7), (1,5), (2,3), (3,3), (4,4), (5,7), (6,0), (6,5), (7,1) **or** (8,1)

Challenge 2
1. **a)** and **b)** Square plotted with vertices at (3,8), (1,6), (5,6) and (3,4)
   c) (3,4)
2. **a)** and **b)** Kite plotted with vertices at (6,6), (4,4), (8,4) and (6,3) **or** (6,6), (4,4), (8,4) and (6,2) **or** (6,6), (4,4), (8,4) and (6,1) **or** (6,6), (4,4), (8,4) and (6,0)
   c) (6,3) **or** (6,2) **or** (6,1) **or** (6,0)

Challenge 3
1. (15,10)
2. (9,9) (accept (−1,4))
3. **a)** (5,5)   **b)** (5,5)
4. No, Daria is not correct. The x-axis displays lines that are vertical. If they are all the same the line would be vertical.

Pages 100–101
Challenge 1
1. **a)** 2   **b)** Red   **c)** Orange
   d) Blue   **e)** 15

Challenge 2
1. **a)** 3   **b)** 38 circled   **c)** 42
   d) 4   **e)** 4

Challenge 3
1. **a)** 19   **b)** 196
   c) 2 second places = 10 points
   4 third places = 8 points
   or
   9 third places = 18 points
   d) Yes, Simon could be correct. Many possible explanations, e.g. 10 second places = 50 points.

Pages 102–103
Challenge 1
1. **a)** 9:00am   **b)** 8°C   **c)** 6°C
   d) 7 hours
   e) 3

Challenge 2
1. **a)** 12:00   **b)** 5 hours
   c) 1 hour
   d) 5°C
   e) Any time between 10:20 and 10:40

Challenge 3
1. **a)** 1°C
   b) 6°C
   c) Any temperature >4°C but <5°C
   d) Any time between 3 hours 30 minutes and 3 hours 45 minutes
   e) Any time between 1 hour 35 minutes and 1 hour 50 minutes

Pages 104–105
Challenge 1
1. **a)** 4   **b)** 28   **c)** 12
   d) 8   **e)** 5

Challenge 2
1. **a)** Sami and Ava   **b)** 100 km
   c) 40 km   **d)** 3

Challenge 3
1. **a)** 60   **b)** 100
   c) 40   **d)** 120

Pages 106–107
Challenge 1
1. **a)** £18   **b)** £32   **c)** £7   **d)** £13

Challenge 2
1. **a)** 5
   b) Hot meal total = 47, added to table
   c) Class A children who had a sandwich = 5, added to table
   d) Class C children who had a vegetarian meal = 9, added to table

Challenge 3
1. **a)** 38
   b) Yes, Olivia is correct. The Friday column shows 24 men visiting. Saturday (18) and Sunday (6) total 24.
   c) 5
   d) Women (accept 76)

Pages 108–111
Progress Test 4
1. 9018, 9108, 9180, 9801, 9810
2. **a)** four hundreds **or** 400
   b) four thousands **or** 4000
   c) four tens **or** 40

Answers

3. 56
4. a) 7503 + 1812 = 9315
   b) 854 − 362 = 492
5. 146
6. $\frac{5}{12}$
7. 2 kg
8. 4100 g, $4\frac{1}{2}$ kg, 4.9 kg, 5 kg, 5500 g
9. 21
10. a) 19:35   b) 20:45   c) 18:50
    (1 mark for all three parts correct)
11. a) 8 kg   b) 9000 ml
    c) 50 cm   d) 450 cm
12. 8.25 litres
13. a) 4.5 kg **or** 4500 g
    b) 4.5 m **or** 450 cm
14. right-angled and scalene circled
15. a) 2   b) 0
16. Obtuse
17. The square has moved **3** squares to the left and **1** square up.
18. a) A
    b) Accept any of the following coordinates: (1,8), (2,7), (7,2), (8,1)
19. a) 155   b) 137   c) 15
20. a) 7 mm
    b) Tuesday and Wednesday
    c) 2
21. a) Kay   b) 8

Notes

Notes

Progress Test Charts

Progress Test 1

Q	Topic	✓ or ✗	See page
1	Counting		12
2	Numbers Beyond 1000		14
3	Numbers Beyond 1000		14
4	Numbers Beyond 1000		14
5	Numbers Beyond 1000		14
6	Place Value		16
7	Place Value		16
8	Roman Numerals		18
9	Roman Numerals		18
10	Representing Numbers		20
11	Representing Numbers		20
12	Rounding Numbers		22
13	Rounding Numbers		22
14	Negative Numbers		24
15	Negative Numbers		24
16	Addition and Subtraction Practice		26
17	Addition and Subtraction Practice		26
18	Estimating and Checking Calculations		28
19	Estimating and Checking Calculations		28
20	Addition and Subtraction Problems		30
21	Addition and Subtraction Problems		30
22	Multiplication and Division Facts		32
23	Multiplication and Division Facts		32
24	Factors		36
25	Factors		36
26	Multiplication Practice		38
27	Multiplication Problems		40
28	Multiplication Problems		40
29	Multiplication Problems		40
30	Multiplication Problems		40

Progress Test 2

Q	Topic	✓ or ✗	See page
1	Counting		12
2	Numbers Beyond 1000		14
3	Place Value		16
4	Representing Numbers		20
5	Rounding Numbers		22
6	Negative Numbers		24
7	Addition and Subtraction Practice		26
8	Estimating and Checking Calculations		28
9	Addition and Subtraction Problems		30
10	Addition and Subtraction Problems		30
11	Factors		36
12	Factors		36
13	Multiplication Practice		38
14	Multiplication Problems		40
15	Multiplication Problems		40
16	Hundredths		46
17	Hundredths		46
18	Equivalent Fractions		48
19	Addition and Subtraction of Fractions		50
20	Fraction and Decimal Equivalents		54
21	Fraction and Decimal Equivalents		54
22	Rounding Decimals		56
23	Rounding Decimals		56
24	Comparing Decimals		58
25	Comparing Decimals		58
26	Dividing by 10 and 100		60
27	Dividing by 10 and 100		60
28	Finding Fractions		52
29	Decimal Problems		62

Progress Test Charts

Progress Test 3

Q	Topic	✓ or ✗	See page
1	Counting		12
2	Numbers Beyond 1000		14
3	Place Value		16
4	Roman Numerals		18
5	Rounding Numbers		22
6	Addition and Subtraction Problems		30
7	Multiplication and Division Facts		32
8	Factors		36
9	Multiplication Practice		38
10	Multiplication Problems		40
11	Hundredths		46
12	Equivalent Fractions		48
13	Addition and Subtraction of Fractions		50
14	Rounding Decimals		56
15	Finding Fractions		52
16	Decimal Problems		62
17	Comparing Measures		68
18	Comparing Measures		68
19	Estimating Measures		70
20	12 and 24 Hour Time		76
21	12 and 24 Hour Time		76
22	Time Problems		78
23	Time Problems		78
24	Converting Measures		72
25	Perimeter and Area		80
26	Perimeter and Area		80

Progress Test 4

Q	Topic	✓ or ✗	See page
1	Numbers Beyond 1000		14
2	Place Value		16
3	Representing Numbers		20
4	Addition and Subtraction Practice		26
5	Multiplication Problems		40
6	Addition and Subtraction of Fractions		50
7	Finding Fractions		52
8	Comparing Measures		68
9	Time Problems		78
10	12 and 24 Hour Time		76
11	Converting Measures		72
12	Measurement Calculations		74
13	Measurement Calculations		74
14	2-D Shapes		86
15	Lines of Symmetry		90
16	Angles		92
17	Translations		94
18	Coordinates / Shapes and Coordinates		96 / 98
19	Tables		106
20	Time Graphs		102
21	Bar Charts		100

What am I doing well in? _____

_____

What do I need to improve? _____

_____